AF890063

Pi, Phi, and The Modern Pagan Calendar

William C. Henry Sr.

Disclaimer

46 St. Books
Published by 46 St. Books
46 St. Books USA Inc.,
Philadelphia, PA 19144, USA

Copyright © William C. Henry Sr., 2015

All rights reserved under International and Pan-American Copyright Conventions.

ISBN 13-9780991652082 ISBN 10-0991652088

Library of Congress Cataloging-in-Publication Data

LCCN#

Henry Sr., William.

PRINTED IN THE UNITED STATES OF AMERICA
Pi, Phi, and the Modern Pagan Calendar

PUBLISHER'S NOTE
This book is a work of non-fictional based on archival scriptures, texts, and documents. All images used are either owned, work for hire, acquired works, use with permission, and from the Library of Congress.

BOOKS ARE AVAILABLE AT QUANTITY DISCOUNTS WHEN USED TO PROMOTE PRODUCTS OR SERVICES. FOR INFORMATION PLEASE WRITE 46 ST. BOOKS, 55 WEST PENN ST, PHILADELPHIA, PA 19144 OR VISIT OUR WEBSITE AT WWW.46STBOOKS.COM

Introduction

I would like to preface this book with a simple and irrefutable fact, that I am in no ways a mathematician! My mind was not designed for numbers and formulas, but was formatted to see the beauty in all things, and the inner connectivity of all systems. I have come to find myself in awe of the math associated with the design of our reality. Science and math tends to rule out our existence as being designed, but their own systems show this to be, as we and our physical reality are no accident. Religion tends to cling to faith in the absence of evidence, even in the face of evidence. When our existence is placed on blind faith, then how can you see all that which is placed before you?

This book is an attempt to show how our physical reality is one that has been misinterpreted accidentally by some, and intentionally by others. When someone understands something fundamental that you don't, then you are prey, as you pray, to their whims. This does not make them smarter, just privy to information that maybe you simply ignored due to personal religious beliefs. Approach all things with an open mind, use your common sense, and never allow your mind or soul to be placed in the box.

Thank you

Table of Contents

TOIFTCOAT

Pi and Phi Basics

Pi is the sixteenth letter in the Greek alphabet. The number π is a mathematic constant that relates the ratio of a circle's circumference to its diameter, and is commonly expressed as 3.14159. It is represented by the Greek letter π, but is also written as Pi. As it is an irrational number, π cannot be expressed exactly as a common fraction, but fractions like 22/7 and other rational numbers are often used to express π. This decimal representation never ends and never settles as a repeating pattern. The transcendental nature of Pi implied that it was impossible to solve the ancient challenge of squaring the circle with a compass and straightedge. Pi is found in many formulae of geometry and trigonometry, as well as cosmology, number theory, statistics, thermodynamics, mechanics, and electromagnetism.

The earliest written approximations of π can be found in Egypt and Babylon. A clay tablet from Babylon from 1900 - 1600 BCE contains a geometrical statement that treats π as 25/8= 3.1250. The Rhind Papyrus from around 1650 BCE was copied from an earlier source document, has a formula for the area of a circle as 3.1605. The Indian Sanskrit text Shulba Sutras from around 600 BCE, treats π as 3.088. In 150 BCE Indian texts gives π a value of 3.1622. Archimedes formatted a geometric polygon around 250 BCE for calculating the value of π.

Phi is the twenty first letter of the Greek alphabet. Phi is also used as a symbol for the golden ratio in math and science. The golden ratio is commonly abbreviated as 1.618, but goes on for infinity such as; 1.6180339887498948420458683 4. The symbol for Phi is Φ as a capital letter, and the lower case form is φ. Phi is used in mathematics, art, and architecture. The recurrence of Phi, has been noted in nature, with the design of the human body, sunflowers, nautilus shells and much more.

The product of Pi and Phi (π X Φ) gives us more clues to the golden circle (Circumference = π) and the golden ellipse (Area = π) of our reality. When 1.61803398 X 3.141592564, we get 5.083203692 as the product of Pi and Phi. This number relates to important tangents points on the golden circle of the sun, and the golden ellipse of our planetary orbit around the sun. These systems give us a firm foundation in the circumference and area of a circle. I will show through this project how they affect our yearly calendar, and have formed our modern Pagan calendar.

The Pythagorean theorem of $a^2+b^2=c^2$ *is also exhibited in Kepler's triangle formula of* $Q=1+\sqrt{5}/2$ $(1+\Phi=\Phi^2)$. *Kepler's triangle is a right triangle with edge lengths in geometric progression. The edges of a Kepler triangle are directly linked to the golden ratio, as there is a ratio between the short side and the hypotenuse equal to the golden ratio. When constructing a Kepler triangle, one must form a golden rectangle.*

This relationship is shown when the circle is squared, as they both have approximately the same perimeter. The Egyptians did understand the Pi and Phi principals, long before Pythagoras learned these principles from them. Modern scholars hold that the Pi and Phi exhibited in the construction of the pyramids is merely a coincidence. There are no coincidences in this existence, simply occurrences!

The Egyptians believed in "As Above So Below," and the pyramids were representations of celestial alignments. A perfect Kepler triangle is formed when squaring the circle of the heavens. The 72° key solar magnetic point with a Pi square root radian of 1.25663706, to the 180° point of Pi radian of 3.14159265, then to the center of the circle forms a triangle 54° x 54° x 72°. When this triangle is halved, we get a right triangle with a geometric progression of 36° x 54° x 90°. This is exactly what we find in the Great Pyramid of Giza, as the sides are concave, and designed to line up with the equinoxes and solstices on the exact dates. The Great Pyramid is located at the center of the land mass of the earth, but again, this is simply mere coincidence!

The heavens form a decahedron when a line is drawn to the important points in the sun phases as noted of; 36° - 72° - 108° - 144° - 180° - 216° - 252° - 288° - 324° - and 360°. Examples; 36 cycled 10 times = 360, 72 solar magnetic point/720 hours in a month (18 hour), rings of Saturn 108,000 miles in diameter, 144,000 days in 1 Baktun, 180° equal triangle sum, 216 is half of the 432 hertz main frequency, 252 solar magnetic point, 2,888,000 days in 1 Piktun, 324,000 hours in a 50 year Jubilee, 360 degrees equals circle or square sum.

Phi /π = 1.1618

Degree	Date	Tangent	Difference
58°	2/28	+1.60033453	58 to 238 = 180
121°	5/1	-1.6642795	
238°	8/28	+1.60033453	
301°	11/1	-1.6642795	121 to 301 =

			180

Pi/Φ = 3.1415

Degree	*Date*	*Tangent*	*Difference*
72°	*3/12*	*+3.07768354*	*72 to 252 = 180*
107°	*4/17*	*-3.2708526*	
252°	*9/12*	*+3.07768354*	
287°	*10/17*	*-3.2708526*	*107 to 287 = 180*

81° - 3/21 - Vernal Equinox

261° - 9/21 - Autumnal Equinox

Pi & Phi/π & Φ Product = 5.0803

Degree	*Date*	*Tangent*	*Difference*
78°	*3/18*	*+4.70463011*	*78 to 258 = 180*
101°	*4/11*	*-5.144554*	
258°	*9/19*	*+4.70463011*	
281°	*10/11*	*-5.144554*	*101 to 281 = 180*

Phi Square Root/Φ^2 *= 1.2720196495*

Degree	*Date*	*Tangent*	*Difference*
51°	*2/21*	*+1.23489716*	*51 to 231 = 180*
128°	*5/8*	*-1.2799416*	
231°	*8/21*	*+1.23489716*	
308°	*11/8*	*-1.2799416*	*128 to 308 = 180*

Pi Square Root/π^2 = 1.77245385091

Degree	*Date*	*Tangent*	*Difference*
60°	*2/30*	*+1.73205081*	*60 to 240 = 180*
120°	*4/30*	*-1.7320508*	
240°	*8/30*	*+1.73205081*	
299°	*10/29*	*-1.8040478*	*120 to 299 = 179*

Pi & Phi Product Square Root/π+Φ √ = 2.25395208467

Degree	*Date*	*Tangent*	*Difference*
66°	*3/**6***	*+2.24603677*	*66 to 246 = 180*
113°	*4/23*	*-2.3558524*	
246°	*9/6*	*+2.24603677*	
294°	*10/24*	*-2.2460368*	*113 to 294 = 181*

360 Day Versus 365 Day Calendar

Our modern 365 day calendar is not the way that things were set up to be. The 360 degrees of the circle equates to our ancient 360 day calendar system. The symmetry between the circle, and the circle of the earth's orbit was disrupted in the recent past. The cosmic disaster that transpired in 2344 BCE, wreaked havoc on the celestial setup as it was designed. The rotation of the planet was slowed, the earth was pushed down in space, and the seasons were altered. Until modern science and religion comes to accept what happened, we will never reconcile the problems that resulted.

The cosmic disaster added 6 hours to the day, as it changed from 18 hours to 24 hours. The axis of the earth was altered from 18 degrees, to 23.5 degrees, as we have a 5.5 degree change. The length of the year went from 360 days, to 365 days, with the addition of 5.5 days, and a leap year added every fourth year. The earth-moon system was altered, as the moon sits at 5.5 degrees above the planet. A solar year went from 400 years, to 394.26 years, a change of 5.74 years. There also happens to be a 5.5 degree crack in the vault of the King's Chamber in the Great Pyramid at Giza. When the synchronicity of the celestial design is fully understood and appreciated, it becomes a wonder of design.

The geometry associated with a circle is another Creative work of wonder, as some important points are the tangent and radian degrees. The Pi (3.1415) and Phi (1.618) aspects of the circle directly affect our reality. Math is simply our system of quantifying Creation in an orderly understanding. I will show through this project how our calendar, math, and religious systems, have been predicated on these ancient understandings.

The aphelion and perihelion points, are the points when the earth is farthest and closest to the sun. These points are standard, and do not change due to precession or any recognized phenomenon. The distance to or from the sun does not affect the seasons, as that is determined by the angle of the planet in regard to the plane of its orbit. We know that our points of aphelion and perihelion changed in the recent past, and we must examine how this happened? The pre 2344 BCE points are listed below, as well as our modern points of aphelion and perihelion.

Perihelion & Aphelion - 2344 BCE or 2012 CE?

*Vernal Equinox 2344 BCE - **3/21** or **81°***

Aphelion 2344 BCE - 4/16 or 106°
Summer Solstice 2344 BCE - ***6/21*** *or* ***171°***
Autumnal Equinox 2344 BCE - ***9/21*** *or* ***261°***
Perihelion 2344 BCE - 10/16 or 286°
Winter Solstice 2344 BCE - ***12/21*** *or* ***351***

Perihelion 2012 CE - 1/3 or 3°
Vernal Equinox 2012 CE - ***3/21*** *or* ***81°***
Summer Solstice 2012 CE - ***6/21*** *or* ***171°***
Aphelion 2012 CE - 7/4 or 184°
Autumnal Equinox 2012 CE - ***9/21*** *or* ***261°***
Winter Solstice 2012 CE - ***12/21*** *or* ***351°***

There is a 77 (perihelion) and 78 (aphelion) day difference between the 2344 BCE and 2012 CE points, for a 77.5 day average change in the aphelion and perihelion of the earth. This change shows how the seasons advanced overnight from April 16th (4/16) of 106° to October 16th (10/16) of 286°. The summer solstice, vernal equinox, winter solstice, and the autumnal equinox, all remained the same. The only thing that could do that to the earth, is a planetary body of comparable size.

Since this cosmic disaster, possibly caused by Venus, humanity tried desperately to reconcile the year. The Bible speaks of renewal of the heavens, and not knowing the proper seasons. The Roman calendar took hundreds of years to settle into a good format. All civilizations had separate calendar years, as the astronomer priests and mathematicians of the time, desperately attempted to reconcile the new orbit of the earth into a proper calendar system. Not until the recent epoch has humanity fixed the modern calendar system, but we still assign Pagan holidays to our modern calendar, and these understandings have shown our amnesia in regard to an ancient disaster.

The 360 degrees of the heavens were divided in 30° portions of 12 each. This lent to a 360 day calendar year, and on most calendars went from the 15th to the 15th of the next month. This same overlapping of months can be attested by the zodiac signs that overlap months. The zodiac on a larger 360° circle is 72 years per degree, and this is called a celestial portion, which is equal to the human lifespan. The heavens were set up with 12 months, 12 zodiac signs, and 12 tribes of Israel (zodiac signs). As Revelations relates the 12 tribes will have 12,000 for each of the 12 tribes. When 12 is multiplied by 12,000, we get 144,000, (days) equal to 400 years when divided by 360, which equals the former tropical year. The King James Version Bible was released in 1611, exactly 400 years before the feared end of the world date!

The Modern Pagan Calendar System

Our modern systems of theology still cling to ancient understandings, that were dispensed to the profane as systems of control. The belief that Gods as planets reigned supreme, and it still does in to modern times. Knowledge was considered only for the enlightened mind, and the lesser minds were given fairy tale knowledge to keep them in place. The manner in which this was done, has translated to a modern ignorance of the reality of nature. There are those with a higher level of true understanding, but it is used for control and manipulation, versus freeing humanity from its self imposed deception.

January 1st to 3rd (1° to 3°) has several joint systems, as the first degree is celebrated as New Years Day (US/Ja/Ru), Consul Start (Ro), Circumcision of Christ and St. Basil Feast (EO), and the Solemnity of Mary (LC). January 3rd (3°) is celebrated as Compitalia (Ro) and the Solemnity of Mary (Ca), and this is where the term, "The third degree," is derived. January 7th (7°) is observed as Christmas (Ru) in Russia. January 13th to the 15th (13° to 15°) begins to show how modern holidays and religious recognitions were sculpted together. We get the Ides (13th) of the month, baptism of Jesus Christ, Kumbh Mela, and Mekar Sankrati.

January 30th (30°) to February 2nd (32°) shows another cluster of understanding. January 30th (30°) is celebrated as the Feast of the Three Hierarchy' in the Eastern Orthodox church. February 1st (31°) is recognized as Dies Natalis in Rome, or a birthday (natal) of the planet annually. The 32° of February 2nd, shows clearly our modern connection to Pagan beliefs. The Pagan-Wiccan holiday of Imbolc, Candlemas (Ca), the presentation of Jesus Christ (Ch), and Ground Hog Day (US). The ground hog is used in modern times to predict or divination of the weather to come in the near future, versus the Pagan use of animal entrails in the past.

February 13th to the 15th (43° to 45°) shows another joint connection. Parentalia fell on the 13th (43°) in Rome, Vali's Blot (P/W) and Saint Valentine's Day (US) all fall on the 14th (44°), and the recognition of the wolf that suckled the founders of Rome is celebrated on the 15th (45°) as Lupercalia. The first Christian Christmas was celebrated at the Shrine of Lupercal in Rome.

February 23rd (53°) is Defender of the Fatherland Day (Ru).

The ancient Roman calendar began in March, and this gives us a clue to other systems and deities. The Ides of March (15th) was the most important time of the calendar year in Rome, and this date has been used with different understandings by religion. Julius Caesar (JC) was murdered on March 15th. Jesus Christ (JC) was created on this day, as he was supposedly born on December 25th (30 other gods as well). The human gestational cycle is 280 days, and if we move backwards from 355 (12/25), we arrive at the 75 degree point of March 15th. The Jewish faithful recognize March 15th, as the day the Babylonians breached the walls in Jerusalem, but this is simply another recognition of March 15th as an annual solar-magnetic point. The ancients believed that God spread his semen at this point across the planet, thereby impregnating Maia (Mary, Gaia), and bringing the rains (April) for fertilizing the crops. The 72 degree of March 12th is a strong positive tangent point, that ushers in the annual change from winter to the vernal equinox.

**72 Degree Events (Pi degree is reached during the day):*

3/12/1496 - Jews expelled from Styria, 3/12/1737 - Galileo's body moved to Santa Croce Church in Florence, 3/12/1799 - Austria declares war on France, 3/12/1856 - U.S. offers 20 dollar gold piece, 3/12/1860 - U.S. Congress accepts Preemption Bill, 3/12/1917 - Stalin arrives in St. Petersburg, 3/12/1917 - Germany sinks the U.S. Algonquin and Wilson issues orders to arm ships, 3/12/1930 - Gandhi begins march against the British Salt Tax, 3/12/1938 - Germany invades Austria, 3/12/1939 - Pope Pius XII is crowned, 3/12/1947 - Truman issues doctrine against Communism, 3/12/1966 - U.S. nuclear test in Nevada, 3/12/1968 - U.S. nuclear test in Nevada, 3/12/1987 - Russian nuclear test

The equinox takes place on March 21st (81°), and is remembered on the Pagan-Wiccan calendar as Eostara, and this is where the Judeo-Christian tradition of Easter comes from. The vernal equinox is clustered with common holidays such as; Solemnity of St. Joseph (Ca) and Quinquatrus (Ro) on 3/19 - 79°, and 3/20 - 80° Spring Equinox (We/Ja). March 22nd (82°)(VC) 1 Death and 1 Dragon (SR).

April 1st (91°) sees another bundling of beliefs, as it is the Feast of Kings holiday (Ch), April

Fool's Day (US), (VC) 1 Creation, (SR) 1 Reed, and Veneralia (Ro). April 1st is seen as the first day of planting season. Is one an April Fool for believing in the Feast of Kings? April 15th (105°) gives an important understanding of squaring the circle of the heavens, as we have the Temple of Jupiter anniversary (Ro) and Vishu (Hi) on the 13th (103°), the Tamil New Year (Hi) and Bishu (Hi) on the 14th (104°), and Fordicidia (Ro), the Crucifixion of Jesus Christ (JC), and Tax Day (US) all falling on the 15th (105°). April 29th is Showa Day (Ja) at 119°.

May 1st (121°) is celebrated cohesively as Beltane (P/W), May Day (US/We/Intl), International Workers Day (Ru), and Sacrifice to Maia (Ro). All of these holidays were enacted in reverence to the process of renewing the earth and soil through phallic symbolism. May 3rd (123°) is Constitution Memorial Day (Ja), May 4th (124°) is Greenery Day (Ja), and May 5th (125°) is Children's Day (Ja).

May 30th (150°) to June 1st (151°) brings us the holidays of the Visitation of Mary (5/31), Anniversaries of the Juno Moneta Temple, Temple of Mars, and the Temple of Tempestates (Ro) on 6/1. June 19th to the 21st (169° to 171°) shows another clustering of holi-days in regard to the solstice. June 19th (169°) has the Solemnity to

Saints Peter & Paul (Ca), Midsummer Festival (P/W), and Temple of Minerva (Ro). June 20th (170°) has the Anniversary of the Temple of Summanus (Ro), and the Summer Solstice (We). The 21st of June (171°) sees Litha (P/W) and the Summer Solstice (P/W) celebrated.

July 4th (184°) is recognized as Independence Day (US), as this is the point when the earth is at aphelion. The Romans celebrated June 5th (185°) as Poplifugia, or "The People's Fight," and this fight or revolution, is what transpired in the United States for independence. This also corresponds to 1 Birth (VC) and 1 Dragon (SR).

August 1st (211°) is celebrated in several systems as; Lughnasada (P/W), Temple of Spes (Ro), and Calan Awst (We), basically in recognition of the middle of growing season, as plants were seen as most potent medicinally at this point of growth. August 11th (221°) is Mountain Day (Ja) in Japan.

September 1st (241°) having just passed a Phi tangent point we celebrate; Jupiter Tonans & Juno Regina (Ro), Krishna Janmashtaami (Hi), and Indiction of the Ecclesiastic Year (EO).

The strong solar-magnetic point of September 12th (252°) sees Nuakhi (Hi) celebrated on this Pi tangent point. This tangent has seen much disaster in recent years with the 9/11 Disasters in New York, Pennsylvania, and Washington, DC. The foundation of the Pentagon in Washington, DC, was laid on 9/11/1941, some 60 years before it was attacked.

**252 Degree Events (Pi is reached during the day):*

9/11/1786 - Convention of Annapolis revises the Articles of Confederation, 9/11/1919 - U.S. marines invade Honduras, 9/11/1922 - British Mandate for Palestine issued, 9/11/1936 - F.D.R. dedicates Boulder Dam, 9/11/1971 - Nikita Kruschev dies at 77, 9/11/1972 - Olympic Games end after 11 die in hostage crisis, 9/11/1986 - Biggest one day Wall Street stock market plunge, 9/11/1990 - George Bush I delivers New World Order speech to Congress, 9/11/1998 - Kenneth Starr delivers report to Congress on President Bill Clinton - 9/11/2001 - World Trade Center hit by planes, plane goes down in Pennsylvania, and a plane strikes the Pentagon 60 years to the day its foundation was laid.

September 23rd (263°) is Autumnal Equinox Day (Ja).

October 1st (271°) sees Fides and Tigillum Sororium (Ro), and Protection of Mary the Mother of God (EO) held on this degree. The October 15th (285°) degree sees a flurry of holidays associated with this tangent, as we have the beginning of the Harvest Season, October Horse (Ro), Winter Finding (W/P), Shemini Atzeret (J) falls on 10/16, and Hajj falls on 10/18.

November 1st (301°) sees another point of Phi, and a cluster of harvest holidays. We get; Nos Galan Gaeaf (We), Winternacht (P/W), and Halloween (US), all on 10/31. On 11/1 we get; All Saints Day (Ca/LC), Samhain (P/W), Ludi Circenses (Ro), and then All Souls Day falls on 11/2 (302°). November 3rd (303°) is Culture Day (Ja) in Japan.

The last holi-day cluster is of course around the winter solstice of December 21st (351°), as we get; Opalia (Ro) 12/19, Yule (P/W) 12/20, and on 12/21; Divalia (Ro), Alean Arthan (We), Pancha Ganapati (Hi), and Mutternacht (P/W). The winter solstice is the shortest day of the year, as the ancients were led to believe that the sun had descended into hell. December 23rd (353°) is the Emperor's Birthday (Ja).

The December 25th (355°) degree finds us at the quoted day of birth for Jesus Christ (JC), known as the Nativity of Christ (Ca/EO/LC) or Christmas (Ch), it is also known as the Birth of Tammuz (J/B) the Sun God, and Dies Natalis Sol Invictus (Ro) or Birth of the Unconquered Sun. This date of birth for the sun or son of the sun (Jesus Christ), is held by over 30 other deities such as; Mithras, Apollo, Dionysus, Krishna, and many other solar figures. It is almost inconceivable to the so-called enlightened mind, that you would be duped by ancient concepts of theology, but the stark reality, is that it has happened! It is also up to you, to free your own mind and soul from that system of confinement.

Moon Based Calendars

Many faith based religions and societies still form calendars that are based on cycles of the moon, versus the use of a solar calendar year. The Jewish, Islamic, Hindu, Buddhist, Babylonian, Assyrian, Levant, Asian, and other calendars were lunisolar in setup. Christianity also bases the celebration of Easter, on the first full moon preceding the Passover. The ancients chose to focus solely on the moon or the sun, as the chief deity for their faiths, or as the object that predicted behaviors on our planet.

It would be supremely ignorant to believe that planets do not bear an influence on our planet, as well as on us. The moon causes the tides on the planet, moving millions of cubic miles of water daily, and the human body is primarily composed of water. We rotate around the sun in an elliptical orbit, our bodies receiving vital nourishment from the rays. I cannot say if these celestial bodies should be worshipped, or simply understood through the mechanics of the solar system setup.

Monthly Pagan Holidays

January

January Holidays Observed:

January 1 - 1° - tangent +0.01745506

(Ro) 1 Kalends in Rome is when Consuls entered office, by vota publica (public vows for the wellbeing of the republic). Festivals were held for the Cult of Aesculapius. Kalends is where we get the modern term "calendar." January was named after the Roman god of "The Doors," known as Janus. (EO) The Feast of Saint Basil and the Circumcision of Jesus Christ is commemorated. (LCa) The Solemnity of Mary is celebrated in the Latin Catholic Church. (US) New Years Day is recognized.

January 3 - 3° - tangent +0.0524778 - Perihelion 2012 (closest to the Sun) day for the planet

(Ro) Compitalia on the Roman calendar was a movable feast (feriae conceptivae). (Ca) The Assumption of Mother is observed, as Mary is taken in to Heaven.

January 5 - 5° - tangent +0.9961947

(Ro) Dies natalis was celebrated as Founding Day on the Roman calendar, and was held at the shrine of Vica Pota on the Velian Hill.

January 6 - 6° - tangent +0.10510424

(Ca/LCa/US) The Epiphany is held on this day, as the Magi visit Jesus Christ, and (EO) Theophany is held.

January 9 - 9° - tangent +0.1583844

(Ro) Agonalia was held in Rome, in honor of the god Janus.

January 11 - 11° - tangent +0.19438031

(Ro) Carmentalia (The Women's Festival) and Juturna (Goddess of Fountains, Wells) were celebrated in Rome.

January 13 - 13° - tangent +0.23086819

(Ro) The Ides of the month in Rome.

January 14 - 14° - tangent +0.249328

(US) Orthodox New Year is recognized. (Hi) Kumbh Mela (pilgrims bathe in sacred river) and Makar Sankrati (harvest festival) are celebrated in the Hindu faith.

January 16 - 16° - tangent +0.28674539

(J) Tu B'ishvat is the Jewish celebration of trees, and those of their country in particular.

January 17 - 17° - tangent +0.30573068

(Lu) Anthony of Egypt is commemorated in the Lutheran church.

January 19 - 19° - tangent +0.34432761

(Lu) The Martyrdom of Henry of Uppsala is held.

January 20 - 20° - tangent +0.40402623

(Hi) Vasant Panchami (coming of Spring) - Fifth day of the waxing Moon of Magh

January 24 - 24° - tangent +0.44522869

(Ro) The festival of Sementivae was celebrated in Rome, a feriae conceptivae of sowing the Earth, and was also celebrated by the Pagi. The Semen in Sementivae shows the belief that a god spread his seed upon the Earth yearly, thus fertilizing it, and enabling the crops to grow.

January 26 - 26° - tangent +0.48773259

(Lu) Saints Timothy, Titus, and Silas are remembered on this date.

January 30 - 30° - tangent +0.57735027

(EO) The Feast of the Three Holy Hierarchy's is held in the Eastern Orthodox Church, as Basil the Great, Gregory the Theologian, and John Chrysostom.

360 DAY	*HOLIDAY*	*DEG*	*TAN*	*Pi*	*Phi*	*Pi & Phi*	*Moon Phases 2015*
1-Jan	New Years Day (US) - Consuls start (Ro) - Circumcision of Christ/St. Basil Feast (EO) - Solemnity of Mary (LC)	1	Pos 0.0175				
2		2	0.0349				
3	***Perihelion* - Solemnity of Mary (Ca) - Compitalia (Ro)**	**3**	0.0524				
4		4	0.0699				Full Moon
5	Dies Natalis (Ro)	5	0.0875				
6	Epiphany (Ca - LC - US) - Theophany (EO)	6	0.1051				
7		7	0.1228				
8		8	0.1405				
9	Agonalia (Ro)	9	0.1584				
10		10	0.1763				
11	Carmentalia (Ro)	11	0.1944				
12		12	0.2126				
13	Ides (Ro) - Baptism Christ (Ca)	13	0.2309				
14	Kumbh Mela - Mekar Sankrati (Hi)	14	0.2493				3rd Quarter
15		15	0.2679				
16	Tu' Bishvet (J)	16	0.2867				

17		17	0.3057				
18		18	0.3249				
19		19	0.3443				
20	Vasant Panchami (Hi)	20	0.3640+				New Moon
21		21	0.3839				
22		22	0.4040+				
23		23	0.4245				
24	Sementivae (Ro)	24	0.4452				
25		25	0.4663				
26		26	0.4877				1st Quarter
27	Dies Natalis (Ro)	27	0.5095				
28		28	0.5317				
29		29	0.5543				
30	Feast of 3 Holy Hierarchys (EO)	30	0.5774				

February

February Holidays Observed:

February 1 - 31° - tangent +0.60086062

(Ro) Dies natalis for the Temple of Juno Sospita (Mother & Queen) with a sacra at the Grove of

Alernus near the Tiber River at the foot of the Palantine Hill in Rome.

February 2 - 32° - tangent +0.62486935

(P/W) Imbolc (St. Brigids/Spring festival) is recognized on the Pagan-Wiccan calendar. (US) Ground Hog day is recognized in the United States of America. (Ca) Candlemas (Jesus presented at temple in Jerusalem) is celebrated by the Catholic faith. (EO) The Lord Jesus Christ is presented in the temple.

February 13 - 43° - tangent +0.93251509

(Ro) The Ides festival of Faunus on the Tiber Island. The Roman also celebrate Parentalia on this day, a commemoration of the ancestors and the dead among families.

February 14 - 44° - tangent +0.96568877

(Lu) Saints Cyril and Methodius are remembered on this day. (P/W) Vali's Blot is celebrated by the Pagan/Wiccan faith. (US) Saint Valentine Day is held for lovers.

February 15 - 45° - tangent +1

(Ro) Lupercalia was celebrated in Rome for the wolf that suckled the founders Romulus and Remus. The

first Christmas was celebrated at the Lupercal Shrine.

February 17 - 47° - tangent +1.07236871

(Ro) Another feriae conceptivae festival Fornicalia, the Oven Festival, and Quirnalia in honor of Quirinus, was celebrated in Rome.

February 21 - 51° - tangent +1.23489716 (Phi square root)

(Ro) Feralia, the only public observation of Parentalia, with dark rituals for the gods below.
(US) Malcolm X assassinated.

February 22 - 52° - tangent +1.27994163

(Ro) Caristia or Dear Kindred, was a family pot luck in the spirit of love and forgiveness in Rome.

February 23 - 53° - tangent +1.32704482

(Ro) In Rome, Terminalia was celebrated in honor of the god Terminus (god who protected boundary markers).

February 24 - 54° - tangent +1.37638192

(Ro) Regifugium was celebrated in Rome.

February 27 - 57° - tangent +1.53986496

(Ro) In Rome the festival of Euiriria held two horse races in honor of the god Mars.

February 30 - 60° - tangent +1.73205081 (Pi square root)

360 DAY	*HOLIDAY*	*DEG*	*TAN*	*Pi*	*Phi*	*Pi & Phi*	*Moon Phases 2015*
1-Feb	Dies Natalis (Ro)	31	0.6009				
2	Imbolc (PW) - Candlemas (Ca) - Christ presented (EO) Ground Hog (US)	32	0.6249				
3		33	0.6494				Full Moon
4		34	0.6745				
5	Dies Natalis (Ro)	35	0.7002				
6		36	0.7265				
7		37	0.7536				
8		38	0.7813				
9		39	0.8098				
10		40	0.8391				
11		41	0.8693				3rd Quarter
12		42	0.9004				
13	Parentalia (Ro)	43	0.9325				
14	Valentine's Day (US) - Vali's Blot (PW)	44	0.9657				
15	Lupercalia (Ro)	45	1				
16		46	1.0355				
17	Fornacalia (Ro)	47	1.0724				

18		48	1.11 06				New Moon
19		49	1.15 04				
20		50	1.19 18				
21	Feralia (Ro)	51	1.23 49		Phi Sq		
22	Caristia (Ro)	52	1.27 99				
23	Terminalia (Ro)	53	1.32 70+				
24	Regifugium (Ro)	54	1.37 64				
25		55	1.42 81				1st Quarter
26		56	1.48 26				
27	Equirria (Ro)	57	1.53 99				
28		58	1.60 03	**1. 6 1 8**			
29		59	1.66 43				
30		60	1.73 21	Pi Sq .			

March

March Holidays Observed:

March 1 - 61° - tangent +1.80404776 (between 2/30 and 3/1 we reach the point of Phi +1.618)

(Ro) March 1st was the original New Years Day on the Roman calendar, when the sacred fire of Rome was renewed. The dancing armed priesthood of

Salii celebrated the Feriae Marti (holiday for Mars). Matronalia was also celebrated in honor of Juno Lucina, the mother of Mars. (US) Saint David's Day is recognized. (Hi) Holi - A full Moon festival that signals Spring. Holi celebrates the slaying of the demoness Holika by Lord Vishnu's (Venus) servant. Holi is where the term Holy or Holi-day derived from.

March 6 - 66° - tangent +2.24603677 (Pi/Phi square root)

March 7 - 67° - tangent +2.35585237

(Ro) In Rome a second festival is held for Vediovis (god that holds arrows or lightning bolts).

March 8 - 68° - tangent +2.47508685

(Int) Flight MH370 goes missing on this degree in 2014.

March 9 - 69° - tangent +2.60508906

(Ro) In Rome, dies religiosus is held with the Salilii carrying the sacred shields around the city. (EO) The Feast of the Forty Martyrs of Sebaste is held.

March 11 - 71° - tangent +2.90421088

(Ja) The Fukushima earthquake occurs, and the nuclear reactor is damaged causing a radioactive leak in 2011. A train is bombed in Madrid in a quoted terrorist attack in 2008.

March 12 - 72° - tangent +3.07768354 (We reach the point of Pi +3.1415 and the radian is the square root of phi 1.2720196495 during that day)

(Lu) Gregory the Great is remembered. (Hi) Maha Shivarati is held by the Hindu faithful. Maha Shivarati is waning Moon festival to Shiva (Venus!) with offerings to Bael (Baal!). This solar point gives the former axis of the earth from 72° to 90° point.

March 14 - 74° - tangent +3.48741444

(Ro) The second Equirria festival is held in Rome, a Feriae Marti called Mamuralia takes place.

March 15 - 75° - tangent +3.73205081 (Pi/Phi square root)

(Ro) This was the most important date in the Roman system, with the Feriae Iovi, sacred to Jove (Jupiter), and also the feast of the year for goddess Anna Perenna. (J/US) Purim is celebrated as the day.

March 16 - 76° - tangent +4.01078093

(Ro) The procession of the Argei is held in Rome.

March 17 - 77° - tangent +4.33147587

(Ro) In Rome, Liberalia is held in honor of Liber, and an Agonalia for Mars is held as well. (Ca/Lu/EO/US) Saint Patrick is celebrated.

March 19 - 79° - tangent +5.14455402 (We reach the product of Pi and Phi between 3/18 and 3/19)

(Ro) Quinquatrus festival is held, and a feast day to Minerva as well, as her temple on Aventine Hill was dedicated on this day. (Ca/LCa) The Solemnity of St. Joseph (husband of Mary/father to Jesus Christ) is observed.

March 20 - 80° - tangent +5.67128182

(We/US) The Spring Equinox is recognized on the Welsh calendar.

March 21 - 81° - tangent +6.31375151 Vernal Equinox 2012 (We reach the sum of double Pi between 3/20 and 3/21)

(P/W) The spring festival of Ostara/Eostara (Dawn goddess associated with rabbits & fertility / Eostre / Easter) takes place.

March 22 - 82° - tangent +7.11536972

(US) The Yale Skull and Bones Club recognizes the Vernal Equinox and Ostara as 322 or 3/22.

March 23 - 83° - tangent +8.14434643

(Ro) Tubilustrium is held in Rome, as purification of the trumpets. (Hi) Gudi Padwa is a waxing moon festival celebrated as New Years day.

March 24 - 84° - tangent +9.51436445

(Ro) The Comita Calata met to sanction wills in Rome. (Hi) Rama Navami celebrates the birth of Rama (Aram, Abram, Brama).

March 25 - 85° - tangent +11.4300523

(EO) Annunciation of the Theotokos (Mary mother of Jesus Christ) is held.

March 30 - 90° - tangent 1.6325E+16 Radian - 1.57079635 (Equals half of Pi)

(Hi) Hanuman Jayanti celebrates the birth of Hanuman (devotee of Rama).

360 DAY	HOLIDAY	DEG	TAN	Pi	Phi	Pi & Phi	Moon Phases 2015
1-Mar	New Years Day (Ro) - Holi (Hi)	61	1.8040+				
2		62	1.8807				
3		63	1.9626				

4		64	2.05 03				
5		65	2.14 45				Full Moon
6		66	2.24 60+			Pi / Phi Sq.	
7	2nd Vediovis Festival (Ro)	67	2.35 59				
8		68	2.47 51				
9	Dies Religiosis (Ro) - Feast of 40 Martyrs of Sebaste (EO)	69	2.60 51				
10		70	2.74 75				
11		71	2.90 42				
12	Maha Shivarati/Bael (Hi)	72	3.07 77		**3. 14 15**		
13		73	3.27 09				3rd Quart er
14	2nd Equirria (Ro)	74	3.48 74				
15	Passover (Ca - Ch) - Purim (J) - ***Conception of Christ (Ch)*** **-** Ides/Feriae Iovi - Julius Caesar death (Ro)	75	3.73 21				
16	Argei (Ro)	76	4.01 08				
17	Liberalia & Agonalia (Ro) - Feast of St. Patrick (EO)	77	4.33 15				
18		78	4.70 46				
19	Solemnity of St. Joseph (Ca - LC) - Ostara (PW) - Quinquatrus (Ro)	79	5.14 46			**5.08 03**	
20	Spring Equinox 20-21 (We)	80	5.67 13				New Moon
21	***Vernal Equinox***	***81***	6.31 38				
22	Skull & Bones 3/22	82	7.11 54				
23	Tubilustrium (Ro) - Gudi Padwa (Hi)	83	8.14 43				

24	QRFC Comita Calata (Ro) - Rama Navami/Ugadi (Hi)	84	9.51 44				
25	Annunciation of Mary (Ca) - Annunciation of Theotokos (EO)	85	11.4 301				
26		86	14.3 007				
27		87	19.0 811				1st Quart er
28		88	28.6 363				
29		89	57.2 900+				
30	3/31 Anniv. Temple of Luna (Ro) - Hanuman Jayanti (Hi)	90	1.62 35E+ 16				

April

April Holidays Observed:

April 1 - 91° - tangent -57.289962 (The Radian on 3/30 - 90° reaches Pi during the course of the day)

(Ro) Veneralia (venereal) is celebrated in Rome, in honor of the goddess Venus. (US) April Fool's Day. (Ch/US) Feast of Kings begins the Easter season. The first day of planting worldwide is 4/1!

April 4- 94° - tangent -14.300666

(Ro) In Rome the Ludi Megalenses is held in honor of Cybele, whose temple was dedicated April 10,

191 BCE. (US) Martin Luther King is assassinated in Memphis, Tennessee in 1968.

April 5 - 95° - tangent -11.430052

(Ro) Anniversary of the Roman Temple of Fortuna Publica.

April 8 - 98° - tangent -7.1153697

(Bu) The birth of Buddha is celebrated in Buddhism.

April 11 - 101° - tangent -5.144554 (We reach the product of Pi and Phi 5.0803 and the radian is the square root of Pi at 1.77245385091 during the course of the day)

April 12 - 102° - tangent -4.7046301

(Ro) Cerialia was held in Rome for Ceres.

April 13 - 103° - tangent -4.3314759

(Ro) Rome celebrates the anniversary of the Temple of Jupiter Victor. (Hi) Vishu is celebrated on the Hindu calendar.

April 14 - 104° - tangent -4.3314759

(Hi) Tamil New Year and Bihu, are observed at this point.

April 15 - 105° - tangent -3.7320508

(Ro) The Roman festival of Fordicidia is held, with the offering of a pregnant cow to Tellus (Earth). (Ca/Ch/US) Easter is celebrated to remember the Crucifixion of Jesus Christ. (J) Yom HaShoah (Holocaust & Hero remembrance day) is held. (US) Tax Day. (P/W) Summerfinding festival is held. (US) Abraham Lincoln is assassinated on this day in 1865. (GB) The R.M.S. Titanic sinks in the Atlantic Ocean after striking an iceberg in 1912.

April 16 - 106° - tangent -3.4874144 Aphelion 2560 BCE

April 17 - 107° - tangent 3.2708526 (We reach the point of Pi 3.1415 during the course of the day)

April 19 - 109° - tangent -2.9042109

(Lu) Olavus Petri (Archbishop of Sweden) and Laurentius Petri (first Lutheran Archbishop of Sweden) are celebrated.

April 21 - 111° - tangent -2.6050891

(Ro) Parilia is held in honor of Pales (deity of shepherds), and the dies natalis in Rome. (Lu) Saint Anselm (Benedictine monk) is commemorated.

April 22 - 112° - tangent -2.4750869

(P/W) Yggdrasil (immense cosmology tree of the nine worlds) is held by the Pagan/Wiccan faith.

April 23 - 113° - tangent -2.3558524 (Pi/Phi square root)

(Ro) The wine festival of Vinalia is held in Rome, in honor of Jupiter, and later for Venus. (EO) The Feast of Saint George (Roman soldier & Christian martyr) is held.

April 24 - 114° - tangent -2.2460368

(US) Arbor Day is recognized in the United States.

April 25 - 115° - tangent -2.1445069

(Ro) The harvest festival of Robigalia is held in Rome with sacrifices of dogs.

April 27 - 117° - tangent -1.9626105

(Ro) In Rome, Ludi Florales is held in honor of Flora (goddess of flowers and spring).

April 30 - 120° - tangent -1.7320508 (Pi square root)

360 DAY	*HOLIDAY*	*DEG*	*TAN*	*Pi*	*Phi*	*Pi & Phi*	*Moon Phases 2015*
1-Apr	Planting Day - Feast of Kings (Ch) - April Fools (US) - Veneralia (Ro)	91	- 57.28 9962				

2		92	- 28.63 6253				
3		93	- 19.08 1137				
4	Ludi Megalenses (Ro)	94	- 14.30 0666				Full Moon
5	Anniv. Of Temple of Fortuna Publica (Ro)	95	- 11.43 0052				
6		96	- 9.514 3645				
7		97	- 8.144 3464				
8	Buddha's Birthday (Bu)	98	- 7.115 3697				
9		99	- 6.313 7515				
10		100	- 5.671 2818				
11		101	- 5.144 554			**5.08 03**	3rd Quarter
12	Cerialia/Ludi Cereri (Ro)	102	- 4.704 6301				
13	Anniv. Of Temple of Jupiter Victor (Ro) - Vishu (Hi)	103	- 4.331 4759				
14	Tamil New Year/Bihu (Hi)	104	- 4.010 7809				
15	Crucifixion (Ch + Ca) - Fordicidia (Ro)	105	- 3.732 0508				
16		106	- 3.487 4144				
17		107	- 3.270		**3.14 15**		

			8526				
18		108	-3.0776835				New Moon
19		109	-2.9042109				
20		110	-2.7474774				
21	Parilia (Ro)	111	-2.6050891				
22	Yggdrasil Day (PW)	112	-2.4750869				
23	Vinalia (Ro) - Feast of St. George (EO)	113	-2.3558524			Pi / Phi Sq.	
24		114	-2.2460368				
25	Robigalia (Ro)	115	-2.1445069				1st Quarter
26		116	-2.0503038				
27	Ludi Florales (Ro)	117	-1.9626105				
28		118	-1.8807265				
29		119	-1.8040478				
30		120	-1.7320508	Pi Sq			

May

May Holidays Observed:

May 1 -121 ° - tangent -1.6642795 (We reach the point of Phi 1.618 during the course of the day)

(P/W) Beltane (spring festival) is celebrated. (US/W) May Day is held. (Ro) A sacrifice to the Earth goddess Maia, anniversary of the Temple of Bona Dea on the Aventine, rites of the Lares Praestites of Rome.

May 2 - 122° - tangent -1.6003345

(Lu) Saint Athanasius is remembered.

May 3 - 123° - tangent -1.539865

(Ro) A celebration for Flora on the anniversary of her temples.

May 4 - 124° - tangent -1.482561

(Lu) Saint Monica is commemorated.

May 5 - 125° - tangent -1.428148

(US) Cinco de Mayo is recognized.

May 8 - 128° - tangent -1.2799416 (Phi square root)

May 9 - 129° - tangent -1.2348972 (radian square root of Pi/Phi product 2.25395208467)

(Ro) Lemuria is held in Rome, as a festival to the dead, with a possible sacrifice to Maia.

May 14 - 134° - tangent -1.0355303

(Ro) Anniversary of the Temple of Mars Invictus (Mars the Unconquered), with a second procession of the Agei in Rome. (US) Ascension Day is held.

May 15 - 135° - tangent -1

(Ro) The ides pass with Mercuralia in honor of Mercury, and a Feriae of Jove.

May 21 - 141° - tangent -0.809784

(Ro) An Agonalia festival is held in honor of Vediovis in Rome. (EO) Feasts of the Holy Emperors Constantine and Helen.

May 23 - 143° - tangent -0.7535541

(Ro) A Tubilustrium Feraie is held for Volcanus (Vulcan) in Rome.

May 25 - 145° - tangent -0.7002075

(Ro) In Rome, the anniversary of the Temple of Primigenia is held. (US) Memorial Day holiday is held.

360 DAY	*HOLIDAY*	*DEG*	*TAN*	*Pi*	*Phi*	*Pi & Phi*	*Moon Phases 2015*
1-May	Beltane (PW) - May Day (US & Intl - We) - Sacrifice to Maia (Ro)	121	-1.6642795		**1.618**		
2		122	-1.6003345				
3	Flora (Ro)	123	-1.539865				Full Moon
4		124	-1.482561				
5		125	-1.428148				
6		126	-1.3763819				
7		127	-1.3270448				
8		128	-1.2799416		Phi Sq.		
9	Lemuria (Ro)	129	-1.2348972				

10		130	- 1.1 917 536					
11		131	- 1.1 503 684				3rd Quarte r	
12		132	- 1.1 106 125					
13		133	- 1.0 723 687					
14	Mars Invictus Temple (Ro)	134	- 1.0 355 303					
15	Mercuralia (Ro)	135	-1					
16		136	- 0.9 656 888					
17		137	- 0.9 325 151					
18		138	- 0.9 004 04				New Moon	
19		139	- 0.8 692 867					
20		140	- 0.8 390 996					
21	Agonalia (Ro) - Feast of Holy Emperors Constantine & Helen (EO)	141	- 0.8 097 84					
22		142	-					

			0.7812856				
23	Tubilustrium (Ro)	143	-0.7535541				
24	QRFC (Ro)	144	-0.7265425				
25	Anniv. Fortuna Primigenia Temple (Ro)	145	-0.7002075				1st Quarter
26		146	-0.6745085				
27		147	-0.6494076				
28	Yerushaylim (J)	148	-0.6248694				
29		149	-0.6008606				
30	May 31 - Visitation of Mary (Ca)	150	-0.5773503				

June

June Holidays Celebrated:

June 1 - 151° - tangent -0.5543091

(Ro) The Roman anniversaries of the Temple of Juno Moneta, of the Temple of Mars, and also the Temple of the Tempestates (storm goddess).

June 3 - 153° - tangent -0.5095254

(Ro) The anniversary of the Temple of Bellona in Rome.

June 4 - 154° - tangent -0.4877326

(Ro) Anniversary of the restoration of the Temple of Hercules Custos in Rome.

June 5 - 155° - tangent -0.4663077

(Ro) Anniversary of the Temple of Dius Fidius in Rome. (Lu) Basil the Great, Gregory of Nyssa, and Gregory of Nazianzus, are remembered.

June 7 - 157° - tangent -0.4244748

(Ro) In Rome, Ludi Piscatorii (Fisherman's Games), and Vestalia, in honor of Vesta are celebrated.

June 8 - 158° - tangent -0.4040262

(Ro) Anniversary of the Temple of Mens in Rome.

June 11 - 161° -tangent -0.3443276

(Ro) The festival of Matralia is celebrated in honor of Mater Matuta is held, and the anniversary of the Temple of Fortuna in the Forum Boarium is held in Rome. (Lu) Benedict of Nursia (patron saint of Europe and students) is remembered.

June 13 - 163° - tangent -0.3057307

(Ro) The Feriae of Jove (Jupiter) is celebrated in Rome, as well as the Quinquatrus minusculae.

June 19 - 169° - tangent -0.2308682

(P/W) The Midsummer Festival occurs in the Paga/Wiccan calendar. (Ro) A commemoration of the Temple of Minerva is held on the Aventine in Rome. (Ca) The Solemnity of Saints Peter and Paul, are observed by the Catholic faith.

June 20 - 170° - tangent -0.2125566

(Ro) The Temple of Summanus anniversary is held in Rome.

June 21 - 171° - tangent -0.1583844

(P/W) Litha (solstice festival) and the June Solstice is celebrated in the Pagan/Wiccan calendar. (We) The Summer Solstice is held. (US) the June Solstice, and Father's Day is held.

June 24 - 174° - tangent -0.1405408

(Ro) A wild affair known as the festival of Fors Fortuna in Rome. (Ca/Lu) The beheading of St. John the Baptist is commemorated. (EO) The Nativity of Saint John is held. (W) Midsummer Day is recognized on the Welsh calendar.

June 27 - 177° - tangent -0.08748887

(Ro) In Rome the anniversary of the Temple of Jupiter Stator is held.

June 29 - 179° - tangent -0.0174551

(Ro) Anniversary of the Temple of Hercules Musarum. (EO) The Feasts of Saints Peter and Paul. (LCa) The Solemnity of Saints Peter and Paul is held.

June 30 - 180° - tangent -1.225E-16 Radian - 3.14159265 is the only point that matches Pi exactly

360 DAY	*HOLIDAY*	*DEG*	*TAN*	*P i*	*P h i*	*Pi & Phi*	*Moon Phases 2015*
1-Jun	Anniv. Of Juno Moneta Temple/Temple of Mars/Temple of Tempestates (Ro)	151	-0.554309 1				
2		152	-0.531709 4				Full Moon
3	Anniv. Of Bellona Temple (Ro)	153	-0.50				

			95254				
4	Aniiv. Of Hercules Custos Temple (Ro)	154	-0.4877326				
5	Anniv. Of Dius Fidius (Ro)	155	-0.4663077				
6		156	-0.4452287				
7	Ludi Piscatorii (Ro) - Vestalia (Ro)	157	-0.4244748				
8	Anniv. Of Mens Temple (Ro)	158	-0.4040262				
9		159	-0.383864				3rd Quarter
10		160	-0.3639702				
11	Matralia/Anniv. Of Fortuna Temple (Ro)	161	-0.3443276				
12		162	-0.3249197				
13	Feriae of Jove/Ides (Ro) - Quinquatrus minusculae (Ro)	163	-0.3057307				
14		164	-0.2867454				
15		165	-0.26				

			79492				
16		166	-0.249328				New Moon
17		167	-0.2308682				
18		168	-0.2125566				
19	Solemnity of St. Peter & Paul (Ca) - Midsummer Festival (PW) - Temple of Minerva (Ro)	169	-0.1943803				
20	Anniv. Of Summanus Temple (Ro) - Summer Solstice 20-21 (We)	170	-0.176327				
21	***Summer Solstice - Litha/June Solstice (PW)***	***171***	-0.1583844				
22		172	-0.1405408				
23		173	-0.1227846				
24	Fors Fortuna (Ro) - St. John Baptist (Ca) - Nativity of St. John (EO) - Midsummer's Day (We)	174	-0.1051042				1st Quarter
25		175	-0.0874887				
26		176	-0.0699268				
27	Lares/Temple of Jupiter Stator (Ro)	177	-0.052407				

			8				
28		178	- 0.03 4920 8				
29	Anniv. Of Hercules Musarum Temple (Ro) - Feast of St. Peter & Paul (EO) - Solemnity St. Peter & Paul (LC)	179	- 0.01 7455 1				
30		180	- 1.25 5 E- 16				**Radian 3.1415**

July

July Holidays Observed:

July 1 - 181° - tangent +0.01745506

(Ro) A Roman anniversary of the Temple to Juno Felicitas.

July 4 - 184° - tangent +0.06992681 - Perihelion 2012 (Earth is farthest from the Sun)

(US) Independence Day is celebrated in the United States.

July 5 - 185° - tangent +0.087448866

(Ro) Poplifugia is celebrated in Rome.

July 6 - 186° - tangent +0.10510424

(Ro) Ludi Appollinares games are held in honor of Apollo in Rome.

July 7 - 187° - tangent +0.12278456

(Ro) Nonae Caprotinae, Ancillarum Feriae (Festival of Serving Women), a sacrifice to Consus by priests, and a minor festival to the two Pales.

July 8 - 188° - tangent +0.14054083

(Ro) The festival of Vitulatio is held in Rome.

July 11 - 191° - tangent +0.19438031

(Lu) Benedict of Nursia is remembered.

July 13 - 193° - tangent +0.23086819

(Hi) Rath Yatra (transport of deities by chariot) is held.

July 14 - 194° - tangent +0.249328

(Ro) A market fair named Mercatus is held in Rome.

July 15 - 195° - tangent +0.26794919

(Ro) Transvectio equitum, a procession of cavalry is celebrated on the Ides.

July 17 - 197° - tangent +0.30573068

(Ro) The anniversary of the Temple of Honos (god of chivalry) and Virtus (deity of virtue) is celebrated in Rome. (J) The holiday the 17th of Tammuz (Sumerian god of food & vegetation) is held. (Int) Flight MH17 is shot down over the Ukraine in 2014.

July 18 - 198° - tangent +0.3249197

(Ro) Dies ater, meaning a black day marks the defeat of Rome by the Gauls.

July 19 - 199° - tangent +34432761

(Ro) The festival of Lucaria is held in Rome.

July 20 - 200° - tangent +0.36397023

(Ro) Ludi Victoriae Caesaris, or the "Games of the Victorious Caesar" is held in Rome. (EO) The Feast of Saint Elijah is held.

July 22 - 202° - tangent +0.40402623

(Ro) Anniversary of the Temple of Concordia is held at the foot of the Capitol.

July 23 - 203° - tangent +0.42447482

(Ro) The Roman festival of Neptunalia is held in honor of Neptune. (Lu) Birgitta of Sweden is commemorated.

July 24 - 204° - tangent +0.44522869

(Ca) St. Mary Magdalene is honored. (EO) The Feast of Saint Christina (Christian martyr) is held.

July 25 - 205° - tangent +0.46630766

(Ro) Furrinalia, feriae publicae is held in honor of Furrina (goddess of water & springs). (Hi) Guru Purnima (dedicated to spiritual & academic teachers) is observed.

July 26 - 206° - tangent +0.48773259

(Lu) Saint Anne the Mother of Mary is remembered.

July 30 - 210° - tangent +0.57735027

(Ro) Anniversary of the Temple of the Fortune of This Day (Fortunae Huiusque Dei) is held in Rome.

360 DAY	*HOLIDAY*	*DEG*	*TAN*	*Pi*	*Phi*	*Pi & Phi*	*Moon Phases 2015*
1-Jul	Anniv. Of Juno Felicitas Temple (Ro)	181	0.017 45508				Full Moon
2		182	0.034 92077				
3		183	0.052 40778				
4	***Aphelion* - Independence Day (US)**	**184**	**0.069 92681**				
5	Poplifugia (Ro)	185	0.087 48866				
6	Ludi Apollinares (Ro) - Anniv. Fortuna Muliebris Temple (Ro)Nonae Caprotinae (Ro)	186	0.105 10424				
7		187	0.122				

			78456				
8	Vitulatio (Ro)	188	0.140 54083				3rd Quarter
9		189	0.158 38444				
10		190	0.176 32698				
11		191	0.194 38031				
12		192	0.212 55656				
13	Rath Yatra (Hi)	193	0.230 86819				
14	Mercatus (Ro)	194	0.249 328				
15	17th of Tammuz (J + Bab) - Transvectio Equitum (Ro)	195	0.267 94919				New Moon
16		196	0.286 74539				
17	Anniv. Honos & Virtus Temple (Ro)	197	0.305 73068				
18	Gaul defeat of Rome (Ro)	198	0.324 9197				
19	Lucaria (Ro)	199	0.344 32761				
20	Ludi Victoriae Caesaris (Ro) - Feast of St. Christina (EO)	200	0.363 97023				
21		201	0.383 86404				
22	Anniv. Concordia Temple (Ro) - St. Mary Magdalene (Ca)	202	0.404 02623				
23	Neptunalia (Ro)	203	0.424 47482				
24		204	0.445 22869				1st Quarter
25	Furrinalia (Ro) - Gudu Purnima (Hi)	205	0.466 30766				
26		206	0.487 73259				
27		207	0.509 52545				
28		208	0.531 70943				
29		209	0.554 30905				

30	Anniv. Fortune of this Day Temple (Ro)	210	0.577 35027				

August

August Holidays Observed:

August 1 - 211° - tangent +0.60086062

(P/W) Lughnasada (harvest festival) is held this day. (R) Anniversary of the Temple of Spes (Hope) in the Forum Holitorium. (W) Calan Awst (harvest festival) is held.

August 3 - 213° - tangent +0.64940759

(Ro) Supplica canum (punishment of the dogs) is held as a weird do sacrifice and procession at the temples of Iuventas (Youth).

August 5 - 215° - tangent +0.70020754

(Ro) A public sacrifice (sacrificium publicum) is held at the Temple of Salus.

August 6 - 216° - tangent +0.72654253 (EO) The Transfiguration of Jesus Christ.

August 9 - 219° - tangent +0.80978403

(Ro) A public sacrifice is held to Sol Indiges (Roman Sun god) in Rome.

August 12 - 222° - tangent +0.90040404

(Ro) A sacrifice of a heifer to Hercules Invictus is held.

August 13 - 223° - tangent +0.93251509

(Ro) The festival of Diana is held on the Aventine (Nemoralia), with Vortumnus, Fortuna Equestris, Hercules Victor, Castor and Pollux, the Camena)Goddess of childbirth), and Flora, being celebrated.

August 15 - 225° - tangent +1

(EO) Domition of the Theotokos is recognized. (LCa/US) Assumption of the Virgin Mary is held.

August 17 - 227° - tangent +1.07236871

(RO) Portunalia is held in honor of Portunes, and the anniversary of the Temple of Janus.

August 19 - 229° - tangent +1.15036841

(Ro) Vinalia Rustica is held in honor of Jupiter, and was later dedicated to Venus.

August 21 - 231° - tangent +1.23489716 (Phi square root)

(Ro) Consualia (harvest festival) is celebrated with a sacrifice on the Aventine in Rome.

August 23 - 233° - tangent +1.32704482

(Ro) Vulcanalia (Feriae Volcano) is celebrated in honor of Vulcan, sacrifices to Maia, the Nymphs in campo, Ops Opifera, and a Hora. (Hi) Raksha Bandhan (love and duty festival) is held today.

August 24 - 234° - tangent +1.37638192

(Ro) Sacrifices to Luna (Moon) on the Graecostasis, and the first three days when the mysterious ritual pit called the mundus opened.

August 25 - 235° - tangent +1.42814801

(Ro) Opiconsivia or Feriae Opi is held in honor of Ops Consivae at the Regia in Rome. (Ca) The Assumption of Mary is held.

August 27 - 237° - tangent +1.53986496

(Ro) Volturnalia celebrated when the Flamen Volturnalis made sacrifices to Volturnus.

August 28 - 238° - tangent +1.60033453 (We reach the Phi 1.618 point between 8/27 and 8/28)

(Ro) The Circus Maximus games are held for Sol (Sun) and Luna (Moon). (Lu) August of Hippo is remembered.

August 29 - 239° - tangent +1.66427948

(EO) John the Baptist beheading.

August 30 - 240° - tangent +1.73205081 (Pi square root)

360 DAY	*HOLIDAY*	*DEG*	*TAN*	*Pi*	*Phi*	*Pi & Phi*	*Moon Phases 2015*
1-Aug	Lughnasada (PW) - Temple of Spes (Ro) - Calan Awst (We)	211	0.600 8606 2				Full Moon 7/31
2		212	0.624 8693 5				
3	Supplica Canum (Ro)	213	0.649 4075 9				
4		214	0.674 5085 2				
5	Sacrificium Publicum (Ro)	215	0.700 2075 4				
6	Transfiguration (EO)	216	0.726 5425 3				3rd Quarter
7		217	0.753 5540 5				
8		218	0.781 2856 3				
9	Sol Indiges (Ro)	219	0.809 7840 3				
10		220	0.839				

			09963				
11		221	0.86928674				
12	Hercules Invictus (Ro)	222	0.90040404				
13	Festival of Diana (Ro)	223	0.93251509				
14		224	0.96568877				New Moon
15	Assumption of Mary (Ca - LC) - Domition of Theotokos/Assumption (EO)	225	1				
16		226	1.03553031				
17	Portunalia/Anniv. Temple of Janus (Ro)	227	1.07236871				
18		228	1.11061251				
19	Vinalia Rustica (Ro)	229	1.15036841				
20		230	1.19175359				
21	Consualia (Ro)	231	1.23489716		Phi Sq.		
22		232	1.27994163				1st Quarter
23	Vulcanalia (Ro) - Raksha Bandhan (Hi)	233	1.32704482				
24	Luna (Ro)	234	1.37638192				
25	Opiconsivia (Ro)	235	1.4281480				

			1				
26		236	1.48256097				
27	Volturnalia (Ro)	237	1.53986496				
28	Circus Maximus (Ro)	238	1.60033453	**1.618**			
29	Beheading of John (EO)	239	1.66427948				Full Moon
30		240	1.73205081	Pi Sq			

September

September Holidays Observed:

September 1 - 241° - tangent +1.80404776

(Ro) Ceremonies for Jupiter Tonans (The Thunderer) on the Capitollium, and Juno Regina on the Aventine. (EO) Indiction of the Ecclesiastical Year. (Hi) Krishna Janmashtami celebrates the birth of Krishna.

September 5 - 245° -tangent +2.14450692

(Ro) Anniversary of the temple of Jupiter Stator, and the festival of Ludi Romani is held.

September 6 - 246° - tangent +Pi/Phi square root)

September 8 - 248° - tangent +2.47508685

(EO) The Nativity of Theotokos is remembered.

September 11 - 251° - tangent +2.90421088

(Hi) Ganesh Chaturthi (remover of obstacles) celebrates Ganesh's arrival on Earth. (US) The World Trade Centers, Pentagon, and Western Pennsylvania, are rocked by four airplanes in a tragic disaster in 2001. The Pentagon has its foundation laid in 1941, exactly 60 years before a plane strikes it.

September 12 - 252° - tangent +3.07768354 (We reach the Pi point +3.1415 during the course of the day)

(Hi) Nuakhi celebrates the new rice of the season on the Hindu calendar.

September 13 - 253° - tangent +3.27085262

(Ro) Anniversary of the Temple to Jupiter Optimus Maximus, an Epulum Iovis, an epulum (feast) to the Capitoline Triad.

September 14 - 254° - tangent +3.48741444

(Ro) Equorum probatio (Approval of the Horses) is held as a cavalry parade. (Ca) Holy Cross day in the

Catholic church. (EO) Elevation of the Cross is held. (J) Rosh Hashanah (New Year) is held.

September 16 - 256° - tangent +4.01078093

(Lu) Saint Cyprian (Bishop of Carthage) is remembered.

September 19 - 259° - tangent +5.14455402 (We reach the product of Pi and Phi 5.0803 between 9/18 and 9/19)

September 20 - 260° - tangent +5.67128182

(Ro) A mercatus is held with markets and fairs in Rome.

September 21 - 261° - tangent +6.31375151

(P/W) The Feast of Mabon (harvest festival) is held.

September 22 - 262° - tangent +7.11536972

(We) The Autumnal Equinox is recognized by the Welsh calendar.

September 23 - 263° - tangent +8.14434643

(R) Anniversary of the rededication of the Temple of Apollo. (J) Yom Kippur is celebrated.

September 25 - 265° - tangent +11.4300523

Rosh Hashanah or Head of the Year, is celebrated in Judaism, and marks the lead up to the Yom Kippur holi-day.

September 26 - 266° - tangent +14.3006663

(Ro) Anniversary of the Temple of Venus Genetrix.

September 30 - 270° - 5.4415E+15 - Radian 4.71238898 (Equals Pi and a half or 3.1415 + 1.5707)

360 DAY	*HOLIDAY*	*DEG*	*TAN*	*Pi*	*Phi*	*Pi & Phi*	*Moon Phases 2015*
1-Sep	Jupiter Tonans/Juno Regina (Ro) - Indiction-Ecclesiastical (EO) - Krishna Janmaashtami (Hi)	241	1.80404776				
2	Anniv. Of Jupiter Stator Temple/Ludi Romani+Ludi Magni (Ro)	242	1.88072647				
3		243	1.96261051				
4		244	2.05030384				
5		245	2.14450692				3rd Quarter
6		246	2.24603677			Pi / Phi Sq.	
7		247	2.35585237				
8	Nativity of Theotokos (EO)	248	2.47508				

			685				
9		249	2.60 508 906				
10		250	2.74 747 742				
11	Ganesh Chaturthi (Hi)	251	2.90 421 088				
12	Nuakhi (Hi)	252	3.07 768 354	3. 14 15			
13	Anniv. Of Jupiter Optimus Maximus Temple (Ro)	253	3.27 085 262				New Moon
14	Equorum Probatio (Ro) - Holy Cross Day (Ca) - Elevation of the Cross (EO)	254	3.48 741 444				
15		255	3.73 205 081				
16		256	4.01 078 093				
17		257	4.33 147 587				
18		258	4.70 463 011				
19		259	5.14 455 402			**5.08 03**	
20		260	5.67 128 182				
21	**Autumnal Equinox - Feast of Mabon (PW)**	**261**	6.31 375 151				1st Quart er
22	Autumn Equinox 22-23 (We)	262	7.11 536 972				
23	Mercatus (Ro)	263	8.14 434 643				

24		264	9.51 436 445				
25	Rosh Hashana (J)	265	11.4 300 523				
26	Anniv. Venus Genatrix Temple (Ro)	266	14.3 006 663				
27		267	19.0 811 367				Full Moon
28		268	28.6 362 533				
29		269	57.2 899 616				
30		270	E - 5.44 15				

October

October Holidays Observed:

October 1 - 271° - tangent -57.2899616

(Ro) Ceremonies for Fides (god of trust) and the Tigillum Sororium (sister's beam). (EO) The protection of the mother of God is held.

October 3 - 273° - tangent -19.081137

(Ro) Ludi Augustales established after the death of Augustus, and is based on Augustalia.

October 4 - 274° - tangent -14.300666

(Ro) A day of fasting titled Ieiunium Cereris in honor of Ceres. (Ca/US) Saint Francis of Assisi is remembered.

October 5 - 275° - tangent -11.430052

(Ro) Second of three days when the mundus pit was opened

October 6 - 276° - tangent -9.5143645

(Ro) Dies ater (Black Day) marks the anniversary of the battle of Arausio.

October 7 - 277° - tangent -8.1443464

(Ro) Rites for Jupiter Fulgur and Juno Curtis.

October 8 - 278° - tangent -7.1153697

(Hi) Navarhthri (deity Durga) is celebrated.

October 9 - 279° - tangent -6.3137515

(Ro) Rites at the shrines for Genius Publicus, Faustus Felicitas, and Venus Victrix on the Capitolium in Rome. (J) In the Jewish faith, Sukkot is known as the Feast of Booths, or the Feasts of Tabernacles. This pilgrimage festival commemorates the time spent in the desert during the Exodus period of disasters.

October 10 - 280° - tangent -5.6712818

(Ro) Rededication of the Temple of Juno Moneta.

October 11 - 281° - tangent -5.144554 (We reach the product of Pi and Phi 5.0803 during the course of the day)

(Ro) The festival of Meditrinalia.

October 12 - 282° - tangent -4.7046301

(Ro) Augustalia is celebrated, and an altar sacrifice to Fortuna Redux.

October 13 - 283° - tangent -4.3314759

(Ro) Fontinalia is held in honor of Fons (god of wells & springs).

October 14 - 284° - tangent -4.0107809

(Ro) Ceremonies mark the restoration of the Temple of the Penates Dei (household deities) on the Velian Hill.

October 15 - 285° - tangent -3.7320508

(Ro) The festival of October Horse with sacrifices to Mars in the Campus Martius, and the Feriae of Jupiter. (Ca) Teresa of Avila (St. Teresa of Jesus) is

celebrated. (P/W) Winterfinding celebration is held. ***Annual Harvest*** *time worldwide.*

October 16 - 286° - tangent -3.4874144 Perihelion 2560 BCE

(J) Shemini Atzeret is a holi-day of Rejoicing with the Torah, features readings from the books of Genesis and Deuteronomy.

October 17 - 287° - tangent -3.2708526 (Pi)

(Hi) Vijayadashami (Rama's victory over Ravana) is celebrated.

October 18 - 288° - tangent -3.0776835

(Is) The festival of Hajj is held. This holiday moves based on lunar cycles.

October 19 - 289° - tangent -2.9042109

(Ro) Armilustrium, a dies religiosus in honor of the god Mars.

October 23 - 293° - tangent -2.3558524

(EO) James the Just is remembered.

October 24 - 294° - tangent -2.2460368 (Pi/Phi square root)

October 26 - 296° - tangent -2.0503038

(Ro) The Victory Games of Sulla. (EO) Saint Demetrius of Thessaloniki is commemorated.

October 29 - 299° - tangent - 1.8040478 (Pi squared)

360 DAY	*HOLIDAY*	*DEG*	*TAN*	*Pi*	*Phi*	*Pi & Phi*	*Moon Phases 2015*
1 Oct	Fides & Tigillum Sororium (Ro) - Protection of the Mother of God (EO)	271	- 57. 289 962				
2		272	- 28. 636 253				
3	Ludi Augustales (Ro)	273	19. 081 137				
4	Ieionium Cereris (Ro) - St. Francis of Assisi (Ca)	274	14. 300 666				3rd Quarte r
5	Mundus (Ro)	275	11. 430 052				
6	Dies Ater (Ro)	276	9.5 143 645				
7	Jupiter Fulgur/Juno Curtis (Ro)	277	8.1 443 464				
8	Genius Publicus/Faleta Felicitas/Venus Victrix (Ro) - Navarhthri (Hi)	278	7.1 153 697				
9	Sukkot (J)	279	6.3 137 515				
10	Temple of Juno Moneta (Ro)	280	- 5.6 712 818				

11	Meditrinalia (Ro)	281	-5.144554			**5.0803**	
12	Augustalia/Fortuna Redux (Ro)	282	-4.7046301				New Moon
13	Fontinalia (Ro)	283	4.3314759				
14	Temple of Penates Dei (Ro)	284	4.0107809				
15	***Harvest Time -*** October Horse (Ro) - Winterfinding (PW)	***285***	-3.7320508				
16	Shemini Atzerat (J)	286	-3.4874144				
17		287	3.2708526	3.1415			
18		288	3.0776835				
19	Armilustrium (Ro)	289	2.9042109				
20		290	2.7474774				1st Quarter
21		291	2.6050891				
22		292	2.4750869				
23	James the Just (EO)	293	2.3558524				
24		294	2.2460368			Pi / Phi Sq.	

25		295	2.1 445 069				
26	Ludi Victoriae Sullanae (Ro) - St. Demetrius of Thessaloniki (EO)	296	2.0 503 038				
27		297	1.9 626 105				Full Moon
28		298	1.8 807 265				
29		299	1.8 040 478	Pi Sq			
30	Nos Galan Gaeaf 31 (We) - Halloween 31 (US)Winternacht (PW)	300	- 1.7 320 508				

November

November Holidays Observed:

November 1- 301° - tangent -1.6642795 (We reach the point of Phi 1.618 during the course of the day)

(P/W) Samhain is celebrated by the Pagan-Wiccan faith. (R) Ludi circenses celebrates an end to the Sullan Victory Games. (Ca/LCa) All Saints Day (All Hallows/Hallowmas) is observed on the Catholic calendar. (We) Nos Galan Gaeaf (first day of winter celebration) is recognized. (US) Halloween falls on 10/31

November 2 - 302° - tangent -1.6003345

(Ca/US) All Souls Day (day of prayer for the dead) is held in the Catholic Church.

November 4 - 304° - tangent -1.482561

(Ro) The Plebeian Games begin.

November 5 - 305° - 1.4281148

(Hi) Deepvali (commemoration of Lord Rama & wife) is celebrated.

November 7 - 307° - -1.3270448

(Hi) Bhai Dooj (sisters pray for brothers) is held.

November 8 - 308° - tangent -1.2799416 (Phi square root)

(Ro) Third day of the opening of the mundus ritual pit. (EO) The Archangels Michael and Gabriel are recognized.

November 11 - 311° - tangent -1.1503684

(P/W) Ancestors Blot (ancestral celebration) and Einherjar (honors kin felled in battle) is held by the Pagan-Wiccan faithful. (Hi) Chhath (Sun God festival) and Prathamastami (life & prosperity ritual).

November 13 - 313° - tangent -1.0723687

(Ro) Epulum Jovis, with ceremonies for Feronia and Fortuna Primigenaie.

November 14 - 314° - tangent -1.0355303

(Ro) An Equorum probatio or cavalry parade is celebrated.

November 18 - 318° - tangent -0.900404

(Ro) A celebration with markets and fairs.

November 21 - 321° - tangent -0.809784

(EO) Entrance of the Theotokos into the temple.

November 22 - 322° - tangent -0.7812856

(US) President John Fitzgerald Kennedy is assassinated in 1963.

360 DAY	*HOLIDAY*	*DEG*	*TAN*	*Pi*	*Phi*	*Pi & Phi*	*Moon Phases 2015*
1-Nov	All Saints Day (Ca - LC) - Samhain (PW) - Ludi Circenses (Ro)	301	- 1.66 427 95		**1.6 18**		
2	All Souls Day (Ca)	302	1.60 033 45				
3		303	1.53 986 5				3rd Quarter

4	Mundus Ritual Pit (Ro)	304	1.48 256 1				
5	Deepvali (Hi)	305	1.42 814 8				
6		306	1.37 638 19				
7	Bhai Dooj (Hi)	307	1.32 704 48				
8	Archangels Michael & Gabriel (EO)	308	1.27 994 16		**Phi Sq.**		
9		309	1.23 489 72				
10		310	1.19 175 36				
11	Einherjar & Ancestors Blot (PW) - Chhath/Prathamastami (Hi)	311	1.15 036 84				New Moon
12		312	1.11 061 25				
13	Epulum Jovis (Ro)	313	1.07 236 87				
14	Equorum Probatio (Ro)	314	1.03 553 03				
15		315	-1				
16		316	0.96 568 88				
17		317	0.93 251 51				
18	Mercars (Ro)	318	1.49 447 65				
19		319	0.86 928				1st Quarter

			67				
20		320	0.83 909 96				
21	Divalia (Ro) - Entrance of Theotokos into the Temple (EO)	321	0.80 978 4				
22		322	0.78 128 56				
23		323	0.75 355 41				
24	***Harvest Thanks***	***324***	0.72 654 25				
25		325	0.70 020 75				Full Moon
26		326	0.67 450 85				
27		327	0.64 940 76				
28		328	0.62 486 94				
29		329	0.60 086 06				
30	Advent (Ca)	330	0.57 735 03				

December

December Holidays Observed:

December 1 - 331° - tangent -0.5543091

(Ro) Ceremonies held at the temples for Neptune and Pietas.

December 3 - 332° - tangent -0.5317094

(Ro) The Bona Dea festival rites for women only.

December 5 - 335° - tangent -0.4663077

(Ro) A festival to the god Faunus (horned god of forest and fields) held by the pagi.

December 6 - 336° - tangent -0.4452287

(EO/US) Saint Nicholas (Nikolas the Wonderworker & secret gift giver, model of Santa Claus) is celebrated.

December 8 - 338° - tangent -0.4040262

(Ro) Festival for Tiberius Pater and Gaia.
(Ca/LCa/US) The Immaculate Conception is celebrated by the Catholic church.

December 9 - 339° - tangent -0.383864

(EO) The Feast of the Conception of Mary is held

December 11 - 341° - tangent -0.3443276

(Ro) Agonalia for Indiges.

December 12 - 342° - tangent -0.3249197

(Ro) Ceremonies at the Temple of Consus on the Aventine. (EO) The Feast of Saint Spiridon is held.

December 13 - 343° - tangent -0.3057307

(Ro) Dies natalis of the Temple of Tellus. (Lu) Saint Lucy is remembered.

December 15 - 345° - tangent -0.2679492

(Ro) Consualia or Feriae for Consus.

December 17 - 347° - tangent -0.2308682

(Ro) Saturnalia in honor of Saturn, with a public ritual.

December 19 - 349° - tangent -0.1943803

(Ro) Opalia in honor of Ops (fertility deity).

December 20 - 350° - tangent -0.176327

(P/W) The Festival of Yule (reformed as Christmas) occurs.

December 21 - 351° - tangent -0.1583844

(We) Alban Arthan is held for the Winter Solstice. (R) Divalia in honor of Angerona, with sacrifices to Hercules and Ceres. (Hi) Pancha Ganapati (lord

Ganesh is celebrated) is observed on the Hindu calendar.

December 22 - 352° - tangent -0.1405408

(Ro) Anniversary of the Temple of Lares Permarini in the Porticus Minucia.(US) December Solstice recognized.

December 23 - 353° - tangent -0.1227846

(Ro) Larentalia commemorations for the temples of Diana and Juno Regina in the Circus Flaminus, and for the Tempestates. Sigillaria celebrated as the last day of Saturnalia (festival to Saturn combined with Christmas) with gift giving.

December 25 - 355° - tangent -0.0874887

(Ro) Dies Natalis Sol Invictus (Birthday of the Unconquered Sun), and Brumalia. (Ca/LCa) The Nativity of Christ is held by the Catholic faithful. (EO) The Nativity of Christ is recognized. (US) Christmas held in the United States.

December 27 - 357° - tangent -0.0524078

(EO)The Feast of Saint Stephen is held.

December 30 - 360° - tangent -2.45E-16 Radian 6.28318531 (Equals Pi doubled)

360 DAY	HOLIDAY	DEG	TAN	Pi	Phi	Pi & Phi	Moon Phases 2015
1-Dec	Temples for Neptune & Pietas (Ro)	331	0.5 543 091				
2		332	0.5 317 094				
3	Bona Dea (Ro)	333	0.5 095 254				3rd Quart er
4		334	0.4 877 326				
5	Faunus (Ro)	335	0.4 663 077				
6	St. Nicholas (EO)	336	0.4 452 287				
7		337	0.4 244 748				
8	Immaculate Conception (Ca - LC) - Tiberius Pater & Gaia (Ro	338	0.4 040 262				
9	Conception of Mary Feast (EO)	339	0.3 838 64				
10		340	0.3 639 702				
11	Agonalia for Indiges (Ro)	341	0.3 443 276				New Moon
12	Temple of Consus (Ro) - St. Spiridon Feast (EO)	342	0.3 249 197				
13	Dies Natalis (Ro)	343	0.3 057 307				
14		344	0.2 867 454				
15	Consualia (Ro)	345	- 0.2				

			679 492					
16		346	0.2 493 28					
17	Hanukkah (J) - Saturnalia (Ro)	347	0.2 308 682					
18		348	0.2 125 566					1st Quart er
19	Opalia (Ro)	349	0.1 943 803					
20	Yule (PW)	350	0.1 763 27					
21	***Winter Solstice - Divalia (Ro) - Alean Arthan 21-22 (We) - Pancha Ganapati (Hi) - Mutternacht (PW)***	***351***	0.1 583 844					
22	Anniv. Lares Permarini Temple (Ro)	352	0.1 405 408					
23		353	0.1 227 846					
24		354	0.1 051 042					
25	Nativity of Christ (Ca-EO - LC) - Christmas (Ch) - Tammuz (J + Bab) - Dies Natalis Sol Invicti (Ro)	355	- 0.0 874 887					Full Moon
26		356	0.0 699 268					
27	St. Stephen Feast (EO)	357	0.0 524 078					
28	Innocent Martyrs (Ca)	358	0.0 349 208					
29		359	0.0 174 551					

30		360	E - 2.4 5					

World Events Based On Solar Tangents

The truth of a system often sits directly before us, but we are often blinded by improper understandings. Major historical events have been aligned with certain solar magnetic tangent points, that coincide with a system of practices and beliefs. We could pick almost any day of the year for historical events, but when you see that the major events fall on consistent tangent points, then another picture of history emerges. I cannot say that the positioning of the events based on tangent points are valid or invalid, but if someone understands something that you do not, then they have a leg up on you!

JANUARY 3RD - Perihelion - 3° tangent +0.05240778
Solemnity of Mary

236 - St Anterus ends his reign as Catholic Pope
269 - St Felix I begins his reign as Catholic Pope
936 - Duke Alberik II of Spoleto appoints his son Pope Leo VII
1431 - Joan of Arc handed over to Bishop Pierre Cauchon
1496 - Leonardo da Vinci unsuccessfully tests a flying machine.
1521 - Martin Luther excommunicated by Pope Leo X from the Roman Catholic Church
1749 - Benning Wentworth issues the first of the New Hampshire Grants, leading to the establishment of Vermont.
1750 - Tax revolt in Haarlem Neth
1752 - East Indies invasion "Geldermalsen" leaves at Malacca: 92 killed
1777 - General George Washington's revolutionary army defeats British forces at Battle of Princeton, NJ

1780 - Danish national anthem "Kong Kristian...," 1st sung
1823 - Stephen F. Austin receives a grant of land in Texas from the government of Mexico.
1825 - Scottish factory owner Robert Owen buys 30,000 acres in Indiana as site for New Harmony utopian community
1831 - 1st US building & loan association organized, Frankford, Penn
1833 - Britain seizes control of Falkland Islands in South Atlantic
1834 - The government of Mexico imprisons Stephen F. Austin in Mexico City.
1840 - 1st deep sea sounding
1848 - Joseph Jenkins Roberts is sworn in as the first President of the independent African Republic of Liberia.
1852 - 1st Chinese arrive in Hawaii
1861 - Delaware legislature rejects proposal to join Confederacy
1861 - US Ft Pulaski & Ft Jackson, Savannah, seized by Georgia. **1862** - Romney Campaign-Stonewall Jackson moves north from Winchester.
1868 - Meiji Restoration returns authority to Japan's emperors
1870 - Construction begins on Brooklyn Bridge in New York; completed May 24, 1883
1871 - Oleomargarine patented by Henry Bradley, Binghamton, NY
1872 - 1st patent list issued by US Patent Office
1888 - 1st wax drinking straw patented, by Marvin C Stone in Washington DC
1889 - Admissions convention meets in Ellensburg, WA, asks for statehood
1890 - 1st US college-level dairy school opens at University of Wisconsin

1896 - Emperor Wilhelm congratulates Pres Kruger on the Jameson Raid

1899 - The first known use of the word automobile, was seen in an editorial in The New York Times.

1900 - Perihelion Passage

1911 - US postal savings bank inaugurated

1911 - The Government of India announces that emigration to Natal, Southern Africa, is prohibited with effect from 1 July

1912 - South Pacific RR offers to bring Liberty Bell to Exposition, free

1918 - US Employment Service opens as a unit of Department of Labor.

1919 - Herbert Hoover is placed in charge of war relief in Europe

1921 - Turkey makes peace with Armenia

1922 - 1st living person identified on a US coin (Thomas E Kirby) on the Alabama Centennial half-dollar

1925 - Benito Mussolini dissolves Italian parliament/becomes dictator

1926 - Greek gen Theodorus Pangulos names himself dictator

1929 - 27 year old William S Paley becomes CBS pres

1932 - Martial law is declared in Honduras to stop revolt by banana workers fired by United Fruit.

1933 - Minnie D. Craig becomes the first female elected as Speaker of the North Dakota House of Representatives, the first female to hold a Speaker position anywhere in the United States.

1938 - March of Dimes established to fight polio.

1939 - Gene Cox becomes 1st girl page in US House of Representatives.

1940 - WPG-AM in Atlantic City NJ consolidates with WBIL & WOV as "new" WOV

1941- Canada & US acquire air bases in Newfoundland (99 yr lease)

1941 - Italian counter offensive in Albania

1941 - Sergei Rachmaninov's "Symphonic Dances" premieres in Philadelphia

1942 - American-British-Dutch-Australian (ABDA) Command forms

1943 - 1st missing persons telecast (NYC)

1943 - Canadian Army troops arrive in North Africa.

1944 - World War II: Top Ace Major Greg "Pappy" Boyington is shot down in his Corsair by Captain Masajiro Kawato flying a Zero.

1945 - Allies land on west coast of Burma, conquer Akyab

1945 - British Premier Winston Churchill visits France

1945 - Greek General Plastiras forms government

1945 - US aircraft carriers attack Okinawa

1946 - As a reward for his wartime cooperation, Governor Thomas E. Dewey commutes Charles "Lucky" Luciano's pandering sentence on condition that he does not resist deportation to Italy

1947 - 1st opening session of Congress to be televised

1947 - William Dawson becomes 1st black to head congressional committee

1948 - Bradman completes dual Test tons (132 & 127*) v India MCG

1949 - "Colgate Theater" dramatic anthology series premieres on NBC TV

1951 - 9 Jewish Kremlin physicians "exposed" as British/US agents

1952 - "Dragnet" with Jack Webb premieres on NBC TV

1953 - Frances Bolton and her son, Oliver from Ohio, become the first mother and son to serve simultaneously in the U.S. Congress. **1955** - Jose Ramon Guizado becomes president of Panama

1958 - The West Indies Federation is formed.

1959 - Alaska admitted as 49th US state
1961 - Adam Clayton Powell elected Chairman of House Education & Labor
1961 - US breaks diplomatic relations with Cuba
1962 - Ground is broken for the Houston Astrodome
1962 - Pope John XXIII excommunicates Fidel Castro
1963 - WOUB TV channel 20 in Athens, OH (PBS) begins broadcasting
1964 - Jack Paar Show, shows a clip of the Beatles singing "She Loves You"
1966 - Floyd B McKissick, named national director of CORE
1967 - "Tonight Show" is shortened from 105 to 90 minutes
1967 - Carl Wilson of the Beach Boys is indicted for draft evasion
1967 - WJAN TV channel 17 in Canton, OH (IND) begins broadcasting
1969 - John Lennon's "2 Virgins" album declared pornographic in NJ
1969 - Rep Adam Clayton Powell Jr seated by Congress
1970 - "Jimmy" closes at Winter Garden Theater NYC after 84 performances
1970 - "Mame" closes at Winter Garden Theater NYC after 1508 performances
1970 - Marxist government takes over in Congo
1970 - WHAG TV channel 25 in Hagerstown, MD (NBC) begins broadcasting
1973 - George Steinbrenner III buys Yankees from CBS for $12 million
1974 - Arias Navarro succeeds Carrero Blanco as premier of Spain
1974 - Burma accepts its constitution
1974 - Gold hits record $121.25 an ounce in London
1976 - US performs nuclear test at Nevada Test Site

1977 - Apple Computer, Inc incorporates

1980 - Gold hits record $634 an ounce

1980 - Babrak Karmal defends the Soviet-backed coup in first public appearance since taking power as President of Afghanistan

1984 - Syria frees captured US pilot after appeal from Jesse Jackson

1985 - Israel government confirms resettlement of 10,000 Ethiopian Jews

1987 - "Oh Coward!" closes at Helen Hayes Theater NYC after 56 performances

1987 - "Smile" closes at Lunt-Fontanne Theater NYC after 48 performances

1987 - Rock & Roll Hall of Fame inducts 1st female artist Aretha Franklin

1988 - Israel orders 9 Palestinian "instigators" deported from W Beirut

1988 - Margaret Thatcher becomes longest-serving British PM this century

1989 - Jim & Tammy Bakker return to TV (Oy Vey!)

1989 - Russian newspaper Izvestia gets its 1st commercial advertisement

1990 - Panama's leader Gen Manuel Noriega surrenders to US authorities

1991 - Israel reopens consulate in USSR after 23 years

1991 - 8 Iraqi embassy officials are expelled from the UK

1993 - "Catskills on Broadway" closes at Lunt-Fontanne NYC after 452 perfs

1993 - "Christmas Carol" closes at Broadhurst Theater NYC after 22 perfs

1993 - "Lost in Yonkers" closes at Richard Rodgers NYC after 780 perfs

1993 - "Secret Garden" closes at St James Theater NYC after

706 performances

1993 - "Tommy Tune Tonite!" closes at Gershwin NYC after 10 performances

1993 - In Moscow, the Start II arms reduction treaty is signed by George Bush (USA) and Boris Yeltsin (Russia)

1994 - "Gray's Anatomy" closes at Beaumont Theater NYC after 13 performances

1994 - 100s killed in Venezuela in prison revolt

1994 - Tupolev-154M crashes at Irkutsk, Siberia: 122 killed

1994 - Restoration of South African citizenship, announced on 15 December 1993 by the South African parliament led by President F.W. de Klerk, becomes effective four months before the first South Africa non-racial polls of 27 April, 1994

1997 - Bryant Gumbel co-hosted his final "Today" show on NBC-TV

1998 - "Side Show," closes at Richard Rodgers NYC after 91 performances

1999 - Israel detains, and later expels, 14 members of Concerned Christians.

1999 - The Mars Polar Lander was launched.

2004 - Flight 604, a Boeing 737 owned by Flash Airlines, an Egyptian airliner, plunges into the Red Sea, killing all 148 people on board.

2004 - After hosting the show for over 30 years, Casey Kasem gives up the hosting duties of "American Top 40" to Ryan Seacrest

2005 - Craig Ferguson becomes the host of "The Late Late Show" on CBS

2007 - National Express has its worst ever coach crash just outside Heathrow Airport.

2009 - Israeli ground forces invade Gaza.

2013 - 27 Shiite pilgrims are killed and 60 are injured by a suicide bombing in Musayyib, Iraq

FEBRUARY 2nd 32° tangent +0.62486935
Ground Hog Day-Candlemas-Imbolc

506 - King Alarik II of Visigoten delegates Lex Romania Visigothorum out
962 - Pope John XII crowns German King Otto I the Great Emperor
1032 - Conrad II, Holy Roman Emperor becomes King of Burgundy, succeeding Rudolf III
1119 - Guido di Borgogna elected Pope Callistus II
1141 - Battle of Lincoln: King Stephen captured by forces loyal to Empress Matilda and commanded by Robert, 1st Earl of Gloucester
1461 - 2nd battle of St Alban's-Lancastrian defeat Yorkists
1536 - Pedro de Mendoza founds Argentine city of Buenos Aires
1542 - Portuguese under Christovão da Gama capture a Moslem-occupied hillfort in northern Ethiopia in the Battle of Baçente.
1550 - Edward Seymour, Duke of Somerset, freed
1633 - M Rossi's opera "Erminia sul Giordano," premieres in Rome
1637 - Zorilla's "El más Impropio Verdugo Para Las," premieres in Madrid
1653 - New Amsterdam becomes a city (later renamed New York)
1709 - British sailor Alexander Selkirk is rescued after being marooned on a desert island for 5 years, his story inspires "Robinson Crusoe"
1714 - Nicholas Rowe's "tragedy of Jane Shore," premieres in London
1731 - George F Handel's opera "Poro," premieres in London
1732 - King Frederik Willem I moves Lutherans towards East-

Prussia

1742 - British Walpole government resigns

1762 - Thomas Arnes opera "Artaxerxes," premieres in London

1787 - Arthur St. Clair is elected the 9th President of the President of the Continental Congress under the Articles of Confederation.

1795 - Joseph Haydns 102nd Symphony in B, premieres

1798 - Federal St Theater, Boston, becomes 1st in US destroyed by fire

1802 - 1st leopard exhibited in US, Boston (admission 25 cents)

1811 - Russian settlers establish Ft Ross trading post, north of SF

1823 - Rossini's opera "Semiramide" premieres in Venice

1829 - Madman Jonathan Martin sets York Cathedral on fire, does £60,000 damage

1843 - US & British settlers in Oregon Country choose government committee

1848 - 1st ship load of Chinese arrive in SF

1848 - Treaty of Guadalupe Hidalgo ends Mexican War; US acquires Texas California, New Mexico & Arizona for $15 million

1852 - 1st British public men's toilet opens (Fleet St London)

1852 - Alexandre Dumas Jr's "Le Dame aux Camélias," premieres in Paris

1854 - Pope Pius IX encyclical "On persecution of Armenians"

1863 - Samuel Clemens becomes Mark Twain for 1st time

1864 - -Oct 7th) Cruise of CSS Florida

1869 - James Oliver invents removable tempered steel plow blade

1870 - Cardiff Giant (supposed petrified human) proved to be gypsum

1878 - Greece declares war on Turkey

1882 - Knights of Columbus forms in New Haven, Conn

1887 - In Punxsutawney, Pennsylvania the first Groundhog Day is observed.

1888 - Frank Sprague opens the first successful U.S. electric street railway system, the Richmond Union Passenger Railway, in Richmond, Virginia

1899 - The Australian Premiers' Conference held in Melbourne decides to locate Australia's capital (Canberra) between Sydney and Melbourne.

1901 - Female Army Nurse Corps established as a permanent organization

1901 - Queen Victoria's funeral takes place.

1901 - Mexican government troops are ambushed by Yaqui Indians, 100 killed

1901 - The US Congress passes the Army Reorganization Act, placing the minimum number of men under arms at 58,000

1906 - Pope encyclical against separation of church & state

1909 - Italian writer Marinetti publishes Futurist Manifest in Paris

1913 - NYC's Grand Central Terminal opens

1914 - James Royce Shannon's musical "Shameen Dhu," premieres in NYC

1919 - Monarchist riot in Portugal

1920 - Estonia declares its Independence from Russia (Dorpat Peace)

1920 - France occupies (German) Memel territory

1920 - Tarto/Dorpat peace treaty: USSR recognizes Estonian independence

1922 - It was 2:22:22 on 2/2/22

1923 - Ethyl gasoline 1st marketed, Dayton, Ohio

1923 - US signs friendship treaty with Central American countries

1925 - Belgian episcopacy rejects liberalism, communism & socialism

1927 - Harry Tierney/Joseph McCarthy's "Rio Rita," premieres in NYC

1927 - Ziegfeld Theater (Loew's Ziegfeld) opens at 6th Ave & 54th St NYC

1931 - 1st siyyum of Talmud celebrated by Daf Yomi students

1931 - 1st use of a rocket to deliver mail (Austria)

1932 - Al Capone sent to prison (Atlanta, Georgia)

1932 - Geneva disarmament conference begins with 60 countries

1932 - Reconstruction Finance Corp organized

1933 - 2 days after becoming chancellor, Adolf Hitler dissolves Parliament

1933 - Hermann Goering bans communist meetings and demonstrations in Germany

1933 - Ucicky's "Rotten Morning," premieres in Berlin

1934 - Dutch RC Bishops warn against fascism/nazism

1935 - Leonarde Keeler first use of his polygraph machine on criminals later convicted of assault on its findings (Portage Wisc)

1942 - LA Times urges security measures against Japanese-Americans

1942 - US auto factories switch from commercial to war production

1943 - Cubs return to original uniform after experimenting with a vest

1943 - German 6th Army surrenders after Battle of Stalingrad, turning point in Europe during WW II

1944 - 4th US marine division conquerors Roi, Marshall Islands

1944 - Allied troops 1st set foot on Japanese territory

1944 - Baseball meets in NYC to discuss postwar action
1944 - Edward Chodorov's "Decision," premieres in NYC
1945 - Escape attempt at Mauthausen concentration camp
1946 - "Nellie Bly" closes at Adelphi Theater NYC after 16 performances
1946 - The Proclamation of Hungarian Republic is made.
1948 - President Harry Truman urges congress to adopt a civil rights program
1949 - Golfing champ Ben Hogan seriously injured in an auto accident
1950 - "Arms & the Girl" opens at 46th St Theater NYC for 134 performances
1950 - 1st broadcast of "What's My Line," on CBS-TV
1951 - -35°F (-37°C), Greensburg, Indiana (state record until 1994)
1951 - US performs nuclear test at Nevada Test Site
1952 - B.B. King's "3 O'Clock Blues" hits #1 on the US Billboard's R&B hit parade to become his first national hit
1954 - Pres Eisenhower reports detonation of 1st H-bomb (done in 1952)
1954 - Snow falls on Gibraltar
1955 - 1st presidential news conference on network TV-Eisenhower on ABC
1956 - Coasters sign with Atlantic Records
1957 - "Candide" closes at Martin Beck Theater NYC after 73 performances
1957 - UN adopts a resolution calling for Israeli troops to leave Egypt
1958 - Fay Crocker wins LPGA Havana Biltmore Golf Open
1958 - Syria joins Egypt in United Arab Republic
1958 - WRIK (now WLUZ) TV channel 7 in Ponce, PR (PTC) begins broadcasting
1962 - 8 of 9 planets align for 1st time in 400 years

1962 - USSR performs nuclear test at Eastern Kazakh/Semipalitinsk USSR

1963 - Helen Shapiro begins tour (Beatles are part of undercard)

1964 - GI Joe, debuts as a popular American boy's toy

1966 - Pakistan suggests a six-point agenda with Kashmir dispute as number one item for the proposed Indo-Pak ministerial talks after 1965 war.

1967 - Bolivia adopts its constitution

1968 - Springer Publishers in West Berlin, bombed

1969 - KMST TV channel 46 in Monterey-Salinas, CA (CBS) begins broadcasting

1971 - Idi Amin ousts Milton Obote to become dictator of Uganda

1972 - Tom Stoppard "Jumpers," premieres in London

1972 - Angry demonstrators burn the British Embassy in Dublin to the ground in protest at the shooting dead of 13 people on 'bloody sunday'

1973 - James R Schlesinger, becomes 9th director of CIA (until July)

1973 - Richath Helms, ends term as 8th director of CIA

1974 - Pope Paul VI encyclical "To Honor Mary"

1974 - The F-16 Fighting Falcon flies for the first time.

1975 - Army offensive against rebels in Eritrea

1975 - Donna Caponi Young wins LPGA Burdine's Golf Invitational

1975 - US female Figure Skating championship won by Dorothy Hamill

1977 - Burn up of Salyut 4 Space Station (USSR)

1977 - Radio Shack officially begins creating TRS-80 computer

1980 - FBI releases details of Abscam, a sting operation that targeted 31 elected & public officials for bribes for political favors

1982 - Government troops and Muslim fundamentalists battle in Hamah, Syria
1983 - Chicago Archbishop Joseph L Bernardin is among 18 new cardinals invested
1983 - Pope John Paul II names 18 new cardinals
1984 - 8th Soap Opera Digest Awards - Days of Our Lives wins
1984 - Lebanese army fight in Beirut
1986 - Dalai Lama meets Pope John Paul II in India
1986 - Oscar Arias Sanchez elected president of Costa Rica
1987 - Philippines adopts constitution
1987 - Islamic militia group kidnap Terry Waite, a Church of England peace envoy in Beirut
1989 - 0°F (-18°C) or below in 15 US states
1989 - FW de Klerk replaces Botha as South Africa's National Party leader
1990 - South Africa's Pres FW de Klerk promises to free Nelson Mandela & legalizes ANC & 60 other political orgs
1992 - IRS & Willie Nelson settle on $9M tax bill (of $16.7M)
1995 - "Moliere Comedies" opens at Criterion Theater NYC for 56 performances
1995 - US space shuttle Discovery launched
1998 - Daniel Baldwin hospitalized in NYC for cocaine overdose
1998 - Philippine DC-9 crashes apparently killing all 104 on board
2012 - Cold snap across Europe kills more than 100 people (over 400 people by 08-02)

MARCH 11th - 71° tangent +2.90421088

2003 - The International Criminal Court holds its inaugural session in The Hague.
2004 - Terrorists explode simultaneous bombs on Madrid's rail network ripping through a commuter train and rocking

three stations, killing 190
2006 - Michelle Bachelet is inaugurated as first female president of Chile.
2009 - Winnenden school shooting - 17 people are killed at a school in Germany.
2010 - Sebastián Piñera become chilean president.
2011 - An earthquake measuring 9.0 in magnitude strikes 130 km (80 miles) east of Sendai, Japan, triggering a tsunami killing thousands of people. This event also triggered the second largest nuclear accident in history, and one of only two events to be classified as a Level 7 on the International Nuclear Event Scale.

March 12th - 72° tangent +3.07768354 Solar Magnetic Difference

538 - Witiges, king of the Ostrogoths, ends his siege of Rome and retreats to Ravenna, leaving the city in the hands of the victorious Roman general, Belisarius.
1054 - Pope Leo IX escapes captivity & returns to Rome
1088 - Odo of Lagery elected as Pope Urban II, succeeding Victor III
1144 - Gherardo Caccianemici elected Pope Lucius II, succeeding Callistus II
1350 - Orvieto city says it will behead & burn Jewish-Christian couples
1365 - University of Vienna founded
1455 - First record of Johann Gutenberg's Bible, letter dated this day by Enea Silvio Piccolomini refers to the bible printed a year before
1496 - Jews are expelled from Syria
1572 - Luis Vaz de Camoes publishes "Os Lusíados" in Portugal
1594 - Company of Distant established for business on East-

Indies
1597 - England routes troops to Amiens
1609 - Bermuda becomes an English colony
1619 - Dutch settlement on Java changes name to Batavia
1622 - Ignatius of Loyola declared a saint
1642 - Abel Tasman is 1st European in New Zealand
1664 - 1st naturalization act in American colonies
1664 - New Jersey becomes a British colony
1689 - Former English King James II lands in Ireland
1737 - Galileo's body moved to Church of Santa Croce in Florence, Italy
1755 - 1st steam engine in America installed, to pump water from a mine
1773 - Jeanne Baptiste Pointe de Sable found settlement now known as Chicago
1794 - Theatre Royal in London's Dury Lane opens after being rebuilt
1799 - Austria declares war on France
1832 - The ballet La Sylphide first premieres at the Opéra de Paris.
1848 - 2nd republic established in France
1849 - 1st gold seekers arrive in Nicaragua en route to Calif
1850 - 1st US $20 gold piece issued
1857 - Giuseppe Verdi's opera "Simon Boccanegra" premieres in Venice
1860 - Congress accepts Pre-emption Bill: free land in West for colonists
1867 - Last French troops leave Mexico
1868 - Great Britain annexes Basutoland in Africa (later renamed the Kingdom of Lesotho)
1868 - US Congress abolishes manufacturer's tax
1868 - Henry O'Farrell attempts to assassinate Prince Alfred, Duke of Edinburgh.

1877 - Great Britain annexes Walvis Bay at Cape colony, Southern Africa.

1881 - Andrew Watson makes his Scotland debut as the world's first black international football player and captain.

1884 - Mississippi establishes 1st US state college for women

1888 - 2nd day of the Great blizzard of '88 in NE US (400 die)

1889 - Battle at Metema (Gallabad): Ethiopian Emperor Yohannes IV, defeated

1894 - Pittsburgh issues free season tickets for ladies on Tuesday & Friday

1894 - In Vicksburg, Mississippi, USA, Coca-Cola is sold in bottles for the first time.

1904 - Andrew Carnegie establishes Carnegie Hero Fund

1905 - The continuing strikes and disorders that unsettle Italy force out Premier Giovanni Giolitti, though he will return in March, 1906

1906 - Heavy storm ravages Dutch west coast

1908 - The Pan-Macedonian group is formed in Athens to support the Greek Struggle for Macedonia.

1909 - Alarmed over increasing German naval strength, Parliament passes a new naval appropriations bill

1913 - Foundation stone of the Australian capital in Canberra laid

1916 - French airship sinks British submarine D3

1917 - [OS Feb 27] Russian Duma sets up the Provisional Committee; Soviets form Executive Committee

1917 - Stalin, Kamenev & Muranov arrives in Petrograd/St Petersburg

1917 - A German submarine sinks an unamred US merchant ship, the 'Algonquin' on the same day that US President Wilson gives executive order to arm US merchant ships

1919 - Austrian National Meeting affirms Anschluss

(incorporate into Germany)

1919 - George Bernard Shaw's "Augustus Does His Bit" premieres in NYC

1925 - British government of Baldwin refuses to ratify Geneva agreement

1926 - Denmark begins unilateral disarmament

1926 - Pope Pius XI names J E van Roey archbishop of Malines Belgium

1928 - In California, the St. Francis Dam fails, killing over 600 people.

1930 - Mohandas Gandhi begins 200m (300km) march protesting British salt tax

1933 - FDR conducts his 1st "fireside chat"

1934 - Acting Pres Constantine Päts commits coup in Tallinn Estonia

1934 - Josip Broz (Tito) freed from jail

1934 - Paul Hindemith's "Mathis der Maler" premieres in Berlin

1935 - England establishes 30 MPH speed limit for towns & villages

1938 - Nazi Germany invades Austria (Anschluss)

1939 - Pope Pius XII crowned in Vatican ceremonies

1940 - Finland surrenders to Russia during WW II, gives Karelische Isthmus

1941 - German occupiers confiscate AVRO studios in Netherlands

1942 - British troops vacate the Andamanen in Gulf of Bengal

1943 - Soviet troops liberate Wjasma

1945 - 30 Amsterdammers executed by nazi occupiers

1945 - Italy's Communist Party (CPI) calls for armed uprising in Italy

1945 - NY is 1st to prohibit discrimination by race & creed in

employment

1945 - USSR returns Transylvania to Romania

1946 - Part of Petsamo province ceded by Soviet Union to Finland

1947 - "Chocolate Soldier" opens at Century Theater NYC for 69 performances

1947 - Pres Harry Truman introduces Truman-doctrine to fight communism

1948 - -5°F lowest temperature ever recorded in Cleveland in March

1950 - Belgium votes (58%) for return of King Leopold III

1950 - Pope Pius XII encyclical "On combating atheistic propaganda"

1951 - Communist troops driven out of Seoul

1954 - 1st performance of Arnold Schoenberg's "Moses und Aaron"

1956 - Dow Jones closes above 500 for 1st time (500.24)

1957 - German DR accepts 22 Russian divisions

1958 - British Empire Day is renamed "Commonwealth Day"

1959 - Dutch Liberal Party wins 2nd parliamentary elections

1959 - US House joins Senate approving Hawaii statehood

1962 - Dutch Premier De Quay announces secret talks with Indonesia

1964 - Jimmy Hoffa sentenced to 8 years

1964 - SN Behrmann's "But for Whom Charlie" premieres in NYC

1964 - WKAB TV channel 32 in Montgomery, AL (ABC) begins broadcasting

1966 - Bobby Hull's 51st goal of season, sets record

1966 - US performs nuclear test at Nevada Test Site

1967 - Indonesian congress deprives president Sukarno of authority

1968 - Mauritius gains independence from Britain (National

Day)

1968 - US performs nuclear test at Nevada Test Site

1969 - 120 cannabis joints found at George & Patti Harrison's home

1970 - US lowers voting age from 21 to 18

1971 - Syrian premier Hafez Assad elected president

1971 - Turkish Government of Demirel forced to resign by Army

1971 - Thousands of Belfast shipyard workers march demanding the introduction of Internment for members of the Irish Republican Army

1972 - Judy Rankin wins LPGA Lady Eve Golf Open

1974 - Bundy victim Donna Manson disappears, Evergreen SC, Olympia, Wash

1975 - Vietcong conquer Ban me Thuot South Vietnam

1976 - South African troops leave Angola

1977 - Chile president Pinochet bans Christian-Democratic Party

1977 - Egypt's Anwar Sadat pledges to regain Arab territory from Israel

1980 - Jury finds John Wayne Gacy guilty of murdering 33 in Chicago

1981 - Soyuz T-4 carries 2 cosmonauts to Salyut 6 space station

1981 - Walter R T Witschey installs world's largest sundial, Richmond, VA

1982 - 1st-class debut of Courtney Walsh, Jamaica v Leeward Islands

1982 - PLO chief Yassar Arafat appears on "Nightline"

1986 - 210.25 million shares traded in NY Stock Exchange

1987 - Federal judge dismisses lawsuits sought by Oliver North

1987 - USSR performs nuclear test at Eastern

Kazakh/Semipalitinsk USSR

1989 - Madagascar AREMA party wins parliamentary election

1991 - OPEC announces oil production cut to 22.3 Mbbl/d (3,550,000 m3/d)

1992 - Mauritius becomes a republic while remaining a member of the Commonwealth of Nations.

1993 - 317 killed by bomb attacks in Bombay

1993 - Cleveland radio station WMMS-FM/101.7 is bought by Disney

1993 - Entertainment Tonight's 3,000th show

1993 - Inkhata leader Mangosuthu Buthelezi begins 2½ week speech

1994 - Church of England ordains 1st 33 women priests

1995 - Congress party loses India national election

1999 - Former Warsaw Pact members the Czech Republic, Hungary and Poland join NATO.

2003 - Zoran Đinđić, Prime Minister of Serbia, is assassinated in Belgrade.

2003 - Elizabeth Smart, was found after having been missing for 9 months.

2004 - Roh Moo-hyun, President of South Korea is impeached by its national assembly for the first time in the nation's history.

2005 - Tung Chee Hwa, the first Chief Executive of Hong Kong, steps down from his post after his resignation is approved by the Chinese central government.

2011 - A reactor at the Fukushima Daiichi Nuclear Power Plant melts and explodes and releases radioactivity into the atmosphere a day after Japan's earthquake.

2012 - 100 people are killed in ethnic clashes and cattle raids in South Sudan

2012 - 45 people, including children, are massacred by the Syrian Army in Homs

2012 - China records its highest trade deficit in over a decade
2013 - JOGMEC becomes the first to successfully extract methane hydrate from seabed deposits
2013 - The 2013 Papal conclave begins with no new Pope elected on the first day
2014 - 8 people are killed, 70 are injured, & 2 buildings are leveled by a gas explosion in East Harlem, New York

MARCH 15th - 75° tangent +3.73205081
Purim/Ides of March/Conception of Christ

1586 - Johan van Oldenbarnevelt becomes Dutch chief legal advisor
1658 - Peace of Roskilde between Sweden & Denmark
1702 - England: Queen Anne ascends throne upon death of King William III
1706 - Vienna's Wiener Stadtbank established
1711 - Antoin de Guiscard tries English premier Haley for murder
1722 - Afghan monarch Mir Mahmud occupies Persia
1746 - Duke of Cumberland's troops occupy Aberdeen
1754 - Marquis of Ensenada becomes premier of Spain
1766 - Willem V (18) becomes governor of United Provinces
1777 - Regiments from Ansbach and Bayreuth, sent to support Great Britain in the American Revolutionary War, mutiny in the town of Ochsenfurt.
1782 - Gnadenhutten Massacre - Ohio militia kills 90 indians
1801 - British drive French forces from Abukir, Egypt
1813 - 1st concerto of Royal Philharmonic
1817 - The New York Stock Exchange is founded.
1838 - US mint in New Orleans begins operation (producing dimes)
1844 - King Oscar I ascends to the throne of Sweden-Norway.
1854 - US Commodore Matthew C Perry's 2nd trip to Japan

1855 - 1st train crosses 1st US railway suspension bridge, Niagara Falls

1861 - St Augustine Florida surrenders to Union armies

1862 - Battle of Elkhorn Tavern ends with Confederate withdrawal

1862 - Confederate ironclad "Merrimack" launched

1862 - Naval Engagement at Hampton Roads, VA CSS Virginia, Jamestown & Yorktown vs USS Cumberland, Congress & Monitor

1865 - Battle of Kingston, NC (Wilcox's ridge, Wise's Forks)

1867 - The British North America Act is passed in the House of Commons, and would serve as a constitution for Canada for the next 100 years

1884 - 1st performance of Edward MacDowell's 2nd Piano suite

1884 - Susan B. Anthony addresses the U.S. House Judiciary Committee arguing for an amendment to the U.S. Constitution granting women the right to vote. Anthony's argument came 16 years after legislators had first introduced a federal women's suffrage amendment.

1887 - Everett Horton, CT, patents fishing rod of telescoping steel tubes

1894 - The state of New York enacts the nation's first dog-licensing law.

1896 - Volunteers of America forms (NYC)

1898 - Richard Straus' "Don Quixote" premieres in Keulen

1908 - Dutch utopist Frederick of Eden speaks in Carnegie Hall, NY

1910 - Baroness Raymonde de Laroche of Paris is 1st licensed female pilot

1911 - International Women's Day is launched in Copenhagen, Denmark, by Clara Zetkin, leader of the Women's Office for the Social Democratic Party in Germany.

1913 - Internal Revenue Service begins to levy & collect income taxes
1915 - 1st US navy minelayer, Baltimore, commissioned
1916 - US invades Cuba for 3rd time, this to end corrupt Menocal regime
1918 - The first case of Spanish flu occurs, the start of a devastating worldwide pandemic.
1920 - **Denmark & Cuba join the League of Nations**
1921 - **Spanish Premier Eduardo Dato Iradier is assassinated while exiting the parliament building in Madrid.**
1924 - Coal mine explosion kills 171 at Castle Gate Utah
1927 - Pan American Airlines incorporates
1929 - US worker union commission reports of slavery in Liberia
1930 - Babe Ruth signs 2-year contract for $160,000 with NY Yankee GM Ed Barrow, wrongly predicts "No one will ever be paid more than Ruth"
1934 - Edwin Hubble photo shows as many galaxies as Milky Way has stars
1942 - **Japanese forces captures Rangoon Burma**
1942 - KNIL, Dutch colonial army on Java, surrenders to Japanese armies
1943 - 335 allied bombers attack German city of Nuremberg, a centre for military production
1943 - Limited gambling legalized in Mexico
1944 - US resumes bombing Berlin
1945 - "Kiss Me Kate" opens in Britain
1945 - 53 Amsterdammers executed by nazi occupiers
1945 - International Women's Day is 1st observed
1945 - Phyllis M Daley is 1st black nurse sworn-in as US Navy ensign
1946 - 1st helicopter licensed for coml use (NYC)
1948 - **US Supreme Court rules in McCollum v. Board of**

Education that relgious instruction in public schools is unconstitutional
1949 - WAGA TV channel 5 in Atlanta, GA (CBS) begins broadcasting
1949 - WBAP-FM, Fort Worth Texas, begins broadcasting
1950 - 1st woman medical officer assigned to naval vessel (BR Walters)
1951 - Intl Table Tennis Federation bans Egypt (for refusing to play Israel)
1952 - Antoine Pinay forms French government
1953 - "Two's Company" closes at Alvin Theater NYC after 90 performances
1953 - Census indicates 239,000 farmers gave up farming in last 2 years
1953 - *KSWO TV channel 7 in Lawton, OK (ABC) begins broadcasting*
1953 - *Patty Berg wins LPGA Jacksonville Golf Open*
1953 - *WFMJ TV channel 21 in Youngstown, OH (NBC) begins broadcasting*
1957 - Israeli troops leave Egypt; Suez Canal re-opened for minor ships
1957 - USSR performs atmospheric nuclear test
1958 - William Faulkner says US school degenerated to become babysitters
1959 - Groucho, Chico & Harpo's final TV appearance together
1959 - *KUAT TV channel 6 in Tucson, AZ (PBS) begins broadcasting*
1959 - Pro-Egyptian coup fails in Mosul Iraq
1960 - "Greenwillow" opens at Alvin Theater NYC for 95 performances
1961 - Jean Kerr's "Mary, Mary" premieres in NYC
1961 - US nuclear submarine Patrick Henry arrives at

Scottish naval base of Holy Loch from SC in a record undersea journey of 66 days 22 hrs

1962 - Beatles, with Pete Best, TV debut (perform "Dream Baby" on BBC)

1962 - US performs nuclear test at Nevada Test Site

1963 - Syrian Arab Rep Revolution Day - Military coup in Syria

1964 - Malcolm X resigns from the Nation of Islam

1965 - 1st US combat forces arrive in Vietnam (3,500 Marines)

1966 - "Golden Boy" closes at Majestic Theater NYC after 569 performances

1966 - An IRA bomb destroyed Nelson's Column in Dublin

1968 - Students demonstrate in Warsaw

***1970** - WTCI TV channel 45 in Chattanooga, TN (PBS) begins broadcasting*

1971 - Joe Frazier beats Muhammad Ali in 15, retains heavyweight boxing title at Madison Sq Garden

1971 - Gun battle between Official Irish Republican Army and Provisional IRA leave 1 man was killed; the feud between the two wings of the IRA had been developing since the split in 1970

1972 - 1st airship flown over Britain in 20 years (Europa)

1973 - Eisenhower Tunnel, world's highest/US longest, opens

1973 - Paul & Linda McCartney are fined £100 for growing cannabis

1973 - US performs nuclear test at Nevada Test Site

1973 - The Provisional Irish Republican Army undertakes its first operation in Great Britain, planting four car bombs in London; 10 members of PIRA are arrested at Heathrow Airport while trying to leave the country

1974 - Charles de Gaulle Airport opens in Paris, France

1975 - Royal Canadian Mint announces branch opening in Winnipeg Manitoba

1976 - 1,774 kg (largest observed) stony meteorite falls in Jilin, China

1978 - The first-ever radio episode of The Hitchhiker's Guide to the Galaxy, by Douglas Adams, is transmitted on BBC Radio 4.

1979 - 1st extraterrestrial volcano discovered (Jupiter's satellite Io)

1979 - China withdraws invasion troops from Vietnam

1979 - Shuttle Columbia (OV-102) transported 38 miles overland from Palmdale

1980 - 6th People's Choice Awards: Burt Reynolds & Jane Fonda win (Motion Picture) and Alan Alda & Carol Burnett win (TV)

1980 - US performs nuclear test at Nevada Test Site

1981 - "Shakespeare's Cabaret" closes at Bijou Theater NYC after 54 perfs

1983 - House Foreign Affairs Com endorses nuclear weapons freeze with USSR

1983 - IBM releases PC DOS version 2.0

1983 - President Reagan calls the USSR an "Evil Empire"

1986 - 4 French TV crew members are abducted in west Beirut Lebanon

1986 - Japanese probe Suisei passes Halley's Comet at 109,800 km

1987 - 17th Easter Seal Telethon raises $35,184,425

1987 - FBI apprehends most wanted Claude L Dallas, Jr in Calif

1987 - "A Team" last aired on NBC-TV after 4 years

1990 - NYC's Zodiac killer shoots 1st victim, Mario Orosco

1991 - US performs nuclear test at Nevada Test Site

1991 - Planeloads of US troops arrive home from the Persian Gulf, Iraq hands over 40 foreign journalists & 2 American soldiers it captured

1992 - 22nd Easter Seal Telethon

1992 - Judy Dickinson wins LPGA Inamori Golf Classic

1994 - B737 collides with Ilyushin-86 in New Dehli, at least 8 killed

1994 - US Defense Department announces smoking ban in workplaces

1994 - Train accident at Pinetown, Natal, kills 47

1995 - Costis Stephanopoulos becomes president of Greece

1995 - Dutch Liberal Party wins Provincial-National elections

1999 - The Supreme Court of the United States upholds the murder convictions of Timothy McVeigh for the Oklahoma City bombing.

2004 - A new constitution is signed by Iraq's Governing Council.

2012 - Greece secures debt-restructuring deal with private lenders

2012 - Toyota recalls 700,000 vehicles over safety concerns

2013 - North Korea terminates all peace pacts with South Korea

2014 - Malaysia Airlines Flight 370 with 239 people loses contact and disappears, prompting the most expensive search effort in history

MARCH 22nd 82° - tangent +7.11536972
Ostara

1984 - Teachers at the McMartin preschool in Manhattan Beach, California are charged with Satanic ritual abuse of the children in the school. The charges are later dropped as completely unfounded.

1985 - *NASA launches Intelsat VA*

1988 - Congress overrides Reagan's veto of sweeping civil rights bill

1989 - US Supreme Court upholds 1 person 1 vote rule of NYC Board of Estimate
1990 - Anchorage jury finds Capt Hazelwood innocent of Valdez oil spill
1991 - NY Daily News begins using motto "Forward with NY"
1991 - Pamela Smart (HS teacher) found guilty in NH of manipulating her student-lover to kill her husband
1992 - US Air NY to Cleveland crashes on take off at LaGuardia, 27 die
1993 - Intel introduces Pentium-processor (80586) 64 bits-60 MHz-100+ MIPS
1994 - South African government/ANC take power in Ciskei homeland
1994 - Soyuz TM-21 lands
1995 - Deputy Gov of Bank of England, Rupert Pennant-Rea, resigns following revelations of his affair with a freelance journalist
1996 - Cheryl Depew, of Florida, crowned 13th Miss Hawaiian Tropic Intl
1996 - STS 76 (Atlantis 16), launches into orbit
1997 - "Sunset Boulevard" closes at Minskoff NYC after 977 performances
1997 - Comet Hale-Bopp Closest Approach to Earth (1.315 AU)
2004 - Ahmed Yassin, co-founder and leader of the Palestinian Sunni Islamist militant group Hamas, and bodyguards are killed in the Gaza Strip when hit by Israeli Air Force AH-64 Apache fired Hellfire missiles.
2006 - ETA, armed Basque separatist group, declares permanent ceasefire.
2006 - BC Ferries' M/V Queen of the North runs aground on Gil Island British Columbia and sinks; 101 on board, 2 presumed deaths.

2006 - Three Christian Peacemaker Teams Hostages are freed by British forces in Baghdad after 118 days captivity and the death of their colleague, American Tom Fox.
2008 - The French Swimmer Alain Bernard sets the world record of 47.50 for the 100 m freestyle long course after winning the European LC Championships 2008.
2009 - Mount Redoubt, a volcano in Alaska began erupting after a prolonged period of unrest.
2011 - Lawrence Taylor pleads guilty for misdemeanors of sexual misconduct and is sentenced to six years probation
2012 - Largest protest in Quebec's history occurs in Montreal with over 200,000 people marching against government tuition hikes and for free access to post-secondary education
2012 - Australia's most wanted man, Malcolm Naden, is captured after seven years on the run in Gloucester, New South Wales
2012 - Massive fire devours thousands of hectares of ancient forests and threatens wildlife on Mount Kenya
2012 - Ireland returns to recession as GDP falls by 0.2% in the fourth quarter of 2011 following a fall of 1.1% in the third quarter
2013 - 37 people are killed and 200 are injured in a refugee camp fire in Ban Mae, Thailand
2014 - 251 people are killed after a boat capsizes in Lake Albert, Uganda
2014 - 30 people are killed by a mudslide in Oso, Washington
2014 - Guinea confirms Ebola outbreak has already killed 59 people
2014 - The US and EU impose sanctions on Russia

JULY 4th 184° - tangent +0.06992681 - Aphelion 2012 (Earth is farthest from the Sun)

836 - Pactum Sicardi, peace between the Principality of Benevento and the Duchy of Naples

993 - Saint Ulrich of Augsburg is canonized.

1054 - Brightest known supernova SN 1054 (creates the Crab Nebula) 1st reported by Chinese astronomers

1120 - Jordan II of Capua is anointed as prince after his infant nephew's death.

1187 - Battle of Hittin (Tiberias): Saladin defeats Reinoud of Châtillon

1301 - Battle at Breukelen: Holland vs Lichtenberg

1359 - Francesco II Ordelaffi of Forlì surrenders to the Papal commander Gil de Albornoz.

1415 - Angelo Correr renounces his claim to the Papacy as Pope Gregory XII

1453 - 41 Jewish martyrs burned at stake at Breslau

1534 - Christian III is elected King of Denmark and Norway in the town of Rye.

1610 - Battle at Klushino: King Sigismund II beats Russian & Sweden

1634 - The city of Trois-Rivières is founded in New France, later to become the Canadian province of Quebec.

1636 - City of Providence, Rhode Island form

1652 - Prince of Condé starts blood bath in Paris

1653 - British Barebones Parliament goes into session

1672 - States of Holland declares "Eternal Edict" void

1693 - Battle at Boussu-lez-Walcourt: French-English vs Dutch army

1708 - Swedish King Karel XII beats Russians

1754 - George Washington gives Ft Necessity to France

1774 - Orangetown Resolutions adopted in the Province of New York, one of many protests against the British Parliament's Coercive Acts

1776 - According to popular legend the Liberty Bell rings for

the 2nd Continental Congress

1776 - US congress proclaims the Declaration of Independence and independence from Britain

1779 - French fleet occupies Grenada

1785 - James Hutton, geologist, publicly reads an abstract of his theory of uniformitarianism for the first time at the meeting of the Royal Society of Edinburgh

1789 - 1st US tariff act

1796 - 1st Independence Day celebration is held

1802 - US Military Academy officially opens (West Point, NY)

1803 - The Louisiana Purchase is announced to the American people.

1810 - French troops occupy Amsterdam

1817 - Construction on Erie Canal begins at Rome, NY

1817 - Chief Engineer James Geddes begins construction on the Erie Canal, one of the first great engineering works in North America

1819 - William Herschel makes last telescopic observation of 1819 comet

1827 - Slavery abolished in NY

1828 - Construction begins on B & O (Baltimore-Ohio) 1st US passenger RR

1829 - Cornerstone laid for 1st US mint (Chestnut & Juniper St, Phila)

1829 - The first London bus "omnibus", operated by George Shillibeer, begins service between Marylebone Road and Bank Junction

1831 - "America (My Country 'Tis of Thee)" is 1st sung in Boston

1836 - Wisconsin Territory forms

1837 - Grand Junction Railway, the world's first long-distance railway, opens between Birmingham and Liverpool.

1838 - Huskar Colliery Mining Disaster in Silkstone England:

Mining pit floods during a rainstorm drowning 26 children, leads to the 1842 commission on the employment of children and women in mines which resulted in the banning of female and child labour underground

1840 - The Cunard Line's 700 ton wooden paddle steamer RMS Britannia departs from Liverpool bound for Halifax, Nova Scotia on the first transatlantic crossing with a scheduled end.

1845 - Henry David Thoreau moves into his shack on Walden Pond

1845 - Texas Congress votes for annexation to US

1855 - In Brooklyn, New York, the first edition of Walt Whitman's book of poems, titled Leaves of Grass, is published.

1861 - In a special session of 27th Congress Lincoln requests 400,000 troops

1861 - Skirmish at Harper's Ferry, WV

1862 - Charles Dodgson (Lewis Carroll) creates Alice in Wonderland for Alice Liddell on a family boat trip on the river Isis (Thames) in Oxford

1862 - Battle of Port Royal, SC (Port Royal Ferry)

1862 - R Morgan's: Tomkinsville, KY to Somerset, KY [->JUL 28]

1863 - Boise, Idaho founded (now capital of Idaho)

1863 - Failed Confederate assault on Helena Arkansas (640 casualties)

1863 - General Lee's army withdraws from Gettysburg

1863 - Skirmish at Smithburg, TN

1863 - Vicksburg, Mississippi surrenders to Union forces

1864 - Battle at Chattahoochee River, Georgia

1865 - First edition of "Alice's Adventures in Wonderland" is published

1866 - Firecracker thrown in wood starts fire destroying half of Portland, Mainee, US

1868 - Battle at Ueno, Japan: last Tokugawa armies defeated
1868 - In New Zealand, Maori leader Te Kooti and 300 of his followers captured the schooner Rifleman and sailed for New Zealand; the fugitives landed at Whareongaonga six days later
1873 - Aquarium opens in Woodward Gardens
1874 - Social Democratic Workmen's Party of North America formed
1875 - White Democrats kill several blacks in terrorist attacks in Vicksburg
1876 - 1st public exhibition of electric light in SF
1876 - Bartholdi visits Bedloe Island, future home of his Statue of Liberty
1879 - Africaner Union forms by Rev SJ du Toit at Cape colony
1879 - Battle at Ulundi: Lord Chelmsford beats Zulu King Cetshwayo
1881 - Booker T. Washington establishes Tuskegee Institute (Alabama)
1882 - Telegraph Hill Observatory opens in SF
1883 - Buffalo Bill Cody presents 1st wild west show, North Platte, Nebr
1884 - 1st US bullfight held (Dodge City Ks)
1884 - Statue of Liberty presented to US in Paris
1886 - 1st scheduled transcontinental passenger train reaches Pt Moody, BC
1887 - The founder of Pakistan, Quaid-i-Azam Muhammad Ali Jinnah, joins Sindh-Madrasa-tul-Islam, Karachi.
1888 - 1st organized rodeo competition held, Prescott, Ariz
1889 - Washington state constitutional convention holds 1st meeting
1892 - James Keir Hardie chosen 1st socialist in British Lower house

1892 - ***Western Samoa changes the International Date Line, so that year there were 367 days in this country, with two occurrences of Monday, July 4.***

1894 - Elwood Haynes successfully tests one of 1st US autos at 6 MPH

1894 - ***Republic of Hawaii proclaimed, Sanford B Dole as president***

1895 - ***Katherine Lee Bates publishes "America the Beautiful"***

1898 - French liner "La Bourgogne" collides with bark Cromartyshire, 560 die

1898 - ***US flag hoisted over Wake Island (Spanish-American War)***

1900 - Williams Jennings Bryan nominated as Democratic Party presidential candidate, USA

1901 - ***William Howard Taft, a former Federal judge, is installed as first governor-general of the Philippines and declares amnesty for all insurgents who take an oath of allegiance***

1902 - ***Civil government is established in the Philippines by a proclamation from US president Roosevelt, who offers a general amnesty to insurgents***

1903 - Pacific Cable (SF, Hawaii, Guam, Phil) opens, Pres TR sends message

1906 - ***Great Britain, France, and Italy declare independence of Ethiopia (Abyssinia), but all lay claim to their own 'spheres of influence' in that land***

1910 - The US Congress pass the Mann-Elkins Act, an important piece of railroad reform legislation

1910 - In a mutual defense agreements, Japan and Russia delineate their spheres of interest in Manchuria

1918 - Altar dedicated at full-scale replica of Stonehenge at Maryhill, Wa

1918 - Ottoman sultan Mehmed VI ascends to the throne.
1919 - ADGB (Allgemeine Deutsche Gewerkschaftsbund) party forms
1920 - The provisional government of Siberia's Maritime Province agrees to hand over parts of the strategic oil- and coal-rich Sakhalin Islands to Japan
1925 - 44 die when Dreyfus Hotel in Boston collapses
1926 - NSDAP-party forms in Weimar
1927 - Ir Sukarno forms PNI (Perserikatan Nasional Indonesia) in Batavia
1927 - First flight of the Lockheed Vega.
1929 - AM radio station WOWO, Indiana's transmitter burns down
1931 - 1st trailside museum opens in Cleveland Metroparks
1933 - Work begins on Oakland Bay Bridge
1934 - Jordanians revolt in Amsterdam after reduction in employment
1934 - Hungarian physicist Leo Szilard patents the chain-reaction design for the atomic bomb.
1936 - League of Nations ends sanctions against Italy after Italian takeover of Abyssinia
1938 - France-Turkish friendship treaty
1940 - British destroys French battle fleet at Oran, Algeria, 1267 die
1940 - German occupiers forbids anti-Nazi speeches
1941 - Latvia partisans shoot 416 Jews dead
1941 - Politburo of Yugoslav communist party reorganizes
1941 - Howard Florey & Norman Heatley meet for 1st time, 11 days later they successfully recreate penicillin
1942 - 1st American bombing mission over enemy-occupied Europe (WW II)
1942 - US air offensive against nazi-Germany begins
1944 - 1,100 US guns fire 4th of July salute at German lines

in Normandy
***1944** - 1st Japanese kamikaze attack, US fleet near Iwo Jima*
***1944** - Allied assault on Carpiquet airport at Caen*
***1944** - Gestapo arrests German Social Democrat Julius Leber*
***1946** - Anti Jewish riots in Kielce Poland, 42 die*
***1946** - Philippines gains independence from US*
***1946** - President Manuel Roxas was inaugurated as the 5th President of the Philippines and the first president of the Third Republic at the Independence Grandstand in Manila*
***1950** - Harry Truman signs public law 600 (Puerto Ricans write own constitution)*
***1950** - The first broadcast by Radio Free Europe.*
***1952** - Canadain Currency, Mint & Exchange Fund Act allows gold coins of $5, $10, and $20 to be minted*
1953 - Imre Nagy succeeds Matyas Rákosi as premier of Hungary
1954 - WMSL (WYUR, now WAFF) TV channel 48 in Huntsville, AL (ABC) begins
1954 - Dr Sam Sheppard's wife Marilyn is murdered (he is accused of crime)
***1956** - Independence National Historical Park forms in Philadelphia*
1956 - US most intense rain fall (1.23" in 1 minute) at Unionville Maryland
1957 - Dutch 2nd Chamber accepts temporary tax increase
***1959** - America's new 49-star flag honoring Alaska statehood unfurled*
***1959** - Cayman Islands separated from Jamaica, made a crown colony*
***1960** - America's new 50-star flag honoring Hawaiian statehood unfurled*
***1961** - Walt Disney is one of the two main speakers on the Independence Day in The Rebuild Hills at Skørping in*

Denmark

1962 - KIKU (now KHNL) TV channel 13 in Honolulu, HI (IND) 1st broadcast

1966 - LBJ signs Freedom of Information Act

1967 - Opening ceremony of Tassajara Zen Mountain Center

1968 - Radio astronomy satellite Explorer 38 launched

1969 - Italian Rumor government resigns

1969 - USSR performs nuclear test at Eastern Kazakh/Semipalitinsk USSR

1969 - The Ohio Fireworks Derecho kills 18 Ohioans and destroys over 100 boats on Lake Erie.

1970 - 100 injured in race rioting in Asbury Park NJ

1970 - Chartered Dan-Air Comet crashes into mountains north of Barcelona, Spain killing 112 vacationing Britons

1970 - The Falls Road curfew in North Ireland, imposed by the British Army while searching for IRA weapons, continues throughout the day; a man is killed by the British Army.

1971 - France performs nuclear test at Muruora Island

1972 - The Royal Ulster Constabulary forward a file about the killings on 'Bloody Sunday' (30 January 1972) to the Director of Public Prosecutions for Northern Ireland

1973 - CARICOM - Caribbean Community & Common Market, forms

1973 - In audience with Italian cyclists, Pope Paul VI praises athletes who "offer the magnificent show of a healthy, strong, generous youth"

1975 - Ted Bundy victim Nancy Baird disappears from Layton, Utah

1976 - Opening ceremony of the Dai Bosatsu monastery Catskill Mt NY

1976 - Operation Entebbe - Israel rescues 229 Air France hostage passengers In Uganda (3 hostages die along with Ugandan soldiers and Israeli soldier)

1978 - Memphis fire fighters halt 3-day strike under a court order

1979 - Algerian ex-president Ben Bella freed

1982 - 4th Space Shuttle Mission-Columbia 4 lands at Edwards AFB

1982 - Miguel de la Madrid Hurtado elected president of Mexico

1982 - USSR performs nuclear test at Eastern Kazakh/Semipalitinsk USSR

1984 - Funeral for S Nakagawa & burial half his ashes next to N Senzaki

1987 - Discovery moves to Launch Pad 39B for STS-26 mission

1987 - Nazi Klaus Barbie, "Butcher of Lyon" sentenced to life in France

1988 - US Navy shoots down Iranian civilian jetliner over Gulf, kills 290

1989 - 14 year old actress Drew Barrymore, attempts suicide

1989 - Unmanned Russian Mig-23 crashes in Bellegem-Kooigem, Belgium (1 dies)

1990 - France performs nuclear test at Muruora Island

1990 - 2 Live Crew release "Banned in the USA" the lyrics quote Star Spangled Banner & Gettysburg Address

1994 - Russian manned space craft TM-18 lands

1994 - Rwandese Patriotic Front occupies Kigali

1996 - Hot Mail, a free internet E-mail service begins

1997 - US space probe Pathfinder lands on Ares Vallis Mars

2003 - Los Angeles Lakers star Kobe Bryant is arrested for sexual assault

2004 - The cornerstone of the Freedom Tower is laid on the site of the World Trade Center in New York City. (This was largely a symbolic event; actual construction would not start for several weeks)

2005 - The Deep Impact collider hits the comet Tempel 1.

2006 - Space Shuttle program: STS-121 Mission - Space Shuttle Discovery launches at 18:37:55 UTC.
2006 - North Korea tests four short-range missiles, one medium-range missile, and a long-range Taepodong-2. The long-range Taepodong-2 reportedly fails in mid-air over the Sea of Japan/East Sea.
2009 - The Statue of Liberty's crown reopens to the public after 8 years, due to security reasons following the World Trade Center attacks.
2013 - 12 people are killed and 60 are injured in a wave of shootings across Chicago
2014 - Rolf Harris is sentenced to 5 years and 9 months for indecently assaulting female minors

JULY 15th 195° tangent +0.26794919
17th of Tammuz

1099 - City of Jerusalem is captured and plundered by Christian forces during the First Crusade
1205 - Pope Innocent III states Jews are doomed to perpetual servitude and subjugation due to crucifixion of Jesus
1207 - King John of England expels Canterbury monks for supporting Archbishop of Canterbury Stephen Langton.
1240 - A Novgorodian army led by Alexander Nevsky defeats the Swedes in the Battle of the Neva.
1307 - Duke Henrik of Corinthia chosen as King of Bohemia
1381 - John Ball, a leader in the Peasants' Revolt, is hung, drawn and quartered in the presence of Richard II of England.
1410 - Battle of Grunwald (2nd Battle of Tannenburg), one of the largest battles of medieval Europe, during Poland-Lithuanian Teutonic War. Polish King Władysław Jagiełło (Jogaila) and Lithuanian Grand Duke Vytautas led defeat against Teutonic Ulrich von Jungingen

1500 - Duke Albrecht of Saxon beats Friese rebellion
1500 - Baglione family massacre at the "Blood Wedding" of Astorre Baglione & Lavinia Colonna in Perugia
1501 - Explorer Pedro Cabral back in Lisbon
1524 - Emperor Karel I bans German national synode
1538 - Peace talks between Karel & King Francois I
1662 - King Charles II charters Royal Society in London
1741 - Alexei Chirikov sights land in Southeast Alaska. He sends men ashore in a longboat, making them the first Europeans to visit Alaska.
1755 - French ambassador recalled from London
1779 - American troops under Gen A Wayne conquer Ft Stony Point, NY
1783 - 1st steamboat, Pyroscaphe, 1st run in France
1787 - Parliament of Paris banished to Troyes
1789 - Gilbert du Motier, marquis de La Fayette, is named by acclamation colonel-general of the new National Guard of Paris.
1795 - "Marseillaise" becomes French national anthem
1799 - The Rosetta Stone is found in the Egyptian village of Rosetta by French Captain Pierre-François Bouchard during Napoleon's Egyptian Campaign.
1806 - Zebulon Pike began his journey to explore the Southwest
1808 - French marshal Joachim Murat becomes king of Naples
1815 - Napoleon surrenders to Captain Frederick Maitland of HMS Bellerophon at Rochefort after his earlier defeat at the Battle of Waterloo
1815 - 1st flat horse race held at English race track Cheltenham on Nottingham Hill
1823 - A fire destroys the ancient Basilica of Saint Paul Outside the Walls in Rome.
1830 - Indian tribes, Sioux, Sauk & Fox, sign fourth Treaty of

Prairie du Chien giving the US most of Minnesota, Iowa & Missouri

1840 - Great Britain, Russia, Austria & Prussia sign Quadruple Alliance

1850 - John Wisden bowls all 10 South batsmen, North v South at Lord's

1856 - Natal forms as a British colony separate from Cape Colony

1862 - CSS Arkansas vs USS Cardondelet & Queen of the West engage at Yazoo R

1863 - Pres Davis orders service duty for confederate army

1864 - Troop train loaded with Confederate prisoners collided with a coal train killing 65 & injuring 109 of 955 aboard

1867 - San Francisco Merchants' Exchange opens

1869 - **Margarine is patented by Hippolye Méga-Mouriès for use by French Navy**

1870 - **Georgia becomes last confederate to be readmitted to US**

1870 - **Manitoba becomes 5th Canadian province & NW Territories created**

1870 - **Hudson's Bay & Northwest Territories transferred to Canada**

1888 - Bandai volcano (Japan) erupts for 1st time in 1,000 years

1904 - **1st Buddhist temple in US forms, Los Angeles**

1906 - Republic museum opens Rembrandt hall in Amsterdam

1911 - 46" of rain (beginning 7/14) falls in Baguio, Philippines

1912 - **British National Health Insurance Act goes into effect**

1914 - Mexican president Huerta flees with 2 million pesos to Europe

1915 - **The head of German propaganda in the US, Dr Heinrich Albert, loses his briefcase on a subway in New York City; an examination of its content reveals an extensive**

network of German espionage and subversion across the US
1915 - **The Austro-German forces launch an offensive along the Eastern Front**
1916 - 22.22" (56.4 cm) of rain falls in Altapass NC (state record)
1916 - **Boeing Company (Pacific Aero) formed by William Boeing in Seattle Wash**
1918 - **World War I: Second Battle of Marne begins**
1922 - 1st duck-billed platypus publicly exhibited in US, at NY zoo
1923 - **Italian parliament accepts new constitution**
1926 - **VPRO (Free thinking Protestant Radio Broadcast) forms**
1927 - Massacre in Vienna: 89 protesters are killed by the Austrian police
1929 - 1st airport hotel opens-Oakland Ca
1932 - **President Hoover cuts own salary 15%**
1933 - **Wiley Post began 1st solo flight around world**
1934 - **Continental Airlines commences operations.**
1936 - Dutch 2nd Chamber agree to temporarily increase defense budget
1937 - **Buchenwald Concentration Camp opens**
1937 - **Japanese attack Marco Polo Bridge, invade China**
1939 - Clara Adams (NYC) is 1st woman to complete round world flight
1940 - **1st betatron placed in operation, Urbana, Il**
1940 - **Nazi occupiers seize library of IISG Amsterdam**
1940 - **Physicist Donald Kerst becomes the first person to accelerate electrons using electromagnetic induction, reaching energies of 2.3 MeV, when his betatron device (for particle acceleration) becomes operational**
1941 - **Florey & Heatley present freeze dried mold cultures (Penicillin)**

1942 - 1st deportation camp at Westerbork, Jews sent to Auschwitz

1942 - Dutch Jews invoked for "Labor camps"

1946 - British North Borneo Company transfers its rights to modern day Sabah to the British Crown

1948 - Alcoholic Anonymous founded in Britain

1948 - Pres Harry Truman nominated for another term (Phila)

1949 - "Miss Liberty" opens at Imperial Theater NYC for 308 performances

1949 *- WBTV TV channel 3 in Charlotte, NC (CBS) begins broadcasting*

1952 - 1st transatlantic helicopter flight begins

1953 - ANC members, Walter Sisulu and Duma Nokwe leave South Africa and go overseas under false names

1954 *- 110°F (43°C) at Balcony Falls, Virginia (state record)*

1954 - 1st coml jet transport plane built in US tested (Boeing 707)

1954 *- KOCO TV channel 5 in Oklahoma City, OK (ABC) begins broadcasting*

1954 *- WBOC TV channel 16 in Salisbury, MD (CBS/NBC/ABC) begins broadcasting*

1955 *- WNDU TV channel 16 in South Bend, IN (NBC) begins broadcasting*

1955 - Eighteen Nobel laureates sign the Mainau Declaration against nuclear weapons, later co-signed by thirty-four others.

1957 - Dutch Super Constellation crashes near New Guinea, 56 die

1957 - US performs nuclear Test at Nevada Test Site

1958 - Pres Eisenhower sends US troops to Lebanon; they stay 3 months

1960 - UN troops arrive in Congo to help deal with the

political crisis after Moïse-Kapenda Tshombé's declaration of independence for Katanga province

1961 - **Spain accepts equal rights for men & women**

1962 - **Algeria becomes member of Arab League**

1962 - Mickey Wright wins LPGA Milwaukee Golf Open

1962 - Netherlands & Indonesia reach agreement over New Guinea

1963 - *KAIT TV channel 8 in Jonesboro, AR (ABC) begins broadcasting*

1964* - *Barry M Goldwater (Sen-R-Az) nominated for president by Republicans

1965* - *"Mariner IV" sends back 1st pictures of Mars

1967 - *USSR performs nuclear Test at Eastern Kazakh/Semipalitinsk USSR*

1968 - "One Life to Live" premieres on TV

1968 - *Commercial air travel begins between US & USSR*

1968 - *France performs nuclear Test at Muruora Island*

1968 - NJ Americans moved to Comack & become NY Nets (ABA)

1968 - Soap opera "One Life To Live" premieres

1971 - **Pres Nixon announces he would visit People's Rep of China**

1973 - Paul Getty III kidnapped

1974 - Military coup on Cyprus: archbishop/president Makarios flees

1975 - Soyuz 19 & Apollo 18 launched; will rendezvous 2 days later

1976 - 36-hr kidnap of 26 school children & their bus driver in Calif

1979 - Morarji Desai resigns as premier of India

1982 - Body of Wendy Caulfield, 1st Green River victim, found near Seattle

1982 - Columbia flies to Kennedy Space Center via Dyess AFB,

Texas

1982 - Senate confirms George Shultz as 60th sec of state by vote of 97-0

1983 - 8 killed, 54 wounded, by Armenian extremists bomb at Orly, France

1987 - Boy George barred from British TV show, he may be a bad influence

1987 - John Poindexter testifies at Iran-Contra hearings

1987 - State of siege ends in Taiwan

1991 - France performs nuclear Test at Muruora Island

1991 - *US troops leave northern Iraq*

1992 - Pope John Paul II hospitalized for 3 weeks to have tumor removed

1992 - The Security Council of the UN examine violence in South Africa

1994 - Gyula Horn sworn in as premier of Hungary

1994 - Israel & Jordan agree to talks in Wash DC on July 25th

1994 - Hundreds of thousands of Hutus flee to Zaire in the Congo near the end of the the Rwandan ***Genocide***

1995 - Jews take Jerusalem

1995 - Northern Virginia begins using new area code 540

1996 - MSNBC begins Microsoft internet-NBC TV

1996 - Southern Mexico hit with 6.5 earthquake

1996 - A Belgian Air Force C-130 Hercules carrying the Royal Netherlands Army marching band crashes on landing at Eindhoven Airport.

1999 - Scientists uncover possible reason for SS Waratah's disappearance 100 years later; The SS Waratah was a luxury steamer which did not reach its destination in 1909, and no trace of the wreckage was ever found

2002 - "American Taliban" John Walker Lindh pleads guilty to supplying aid to the enemy and to possession of explosives during the commission of a felony.

2002* - *Anti-Terrorism Court of Pakistan hands down the death sentence to British born Ahmed Omar Saeed Sheikh and life terms to three others suspected of murdering Wall Street Journal reporter Daniel Pearl.
2003* - *AOL Time Warner disbands Netscape Communications Corporation. The Mozilla Foundation is established on the same day.
2009 - A 7.9 Magnitude earthquake registers 160km west of Invercargill, New Zealand, creating a small tsunami.
2012 - 39 pilgrims are killed in a bus crash in Parasi, Nepal
2012 - A Russian Soyuz rocket with an international team launches for a mission to the International Space Station
2013 - 13 people are killed by a car bomb in Deir Atiyah, Syria
2013 - 18 people are killed and 47 are injured in a riot following a boxing match in Indonesia
2014 - 20 people are killed & 100 are injured after a train derails in Moscow
2014* - *Israel announces it has officially voted to accept the ceasefire proposed by Egypt; Hamas reject, saying they were not consulted

SEPTEMBER 11th ***251° tangent +2.90421088***

9 - The Battle of the Teutoburg Forest ends.
506 - The bishops of Visigothic Gaul meet in the Council of Agde.
813 - Charles the Great crowns Louis I emperor
1185 - Isaac II Angelus kills Stephanus Hagiochristophorites and then appeals to the people, resulting in the revolt that deposes Andronicus I Comnenus and places Isaac on the throne of the Byzantine Empire.
1226 - The Roman Catholic practice of public adoration of the Blessed Sacrament outside of Mass spreads from monasteries to parishes.

1297 - Battle at Stirling Bridge, Scottish rebel William Wallace beats English

1304 - Willem III becomes earl of Holland

1390 - Lithuanian Civil War (1389-1392): the Teutonic Knights begin a five-week siege of Vilnius.

1541 - Santiago, Chile, is destroyed by indigenous warriors, lead by Michimalonko.

1557 - Catholic & Lutheran theology debated in Worms

1609 - Expulsion order announced against the Moriscos of Valencia; beginning of the expulsion of all Spain's Moriscos.

1645 - Thomas Fairfax's 'New Model Army' occupies Bristol

1648 -13] Battle at Pilawce: Bohdan Chmielricki's beats John Casimir [OS]

1649 - Massacre of Drogheda, Ireland - Oliver Cromwell kills 3,000 royalists

1697 - Battle of Zenta: Prince Eugen of Savoye beats Turks, ending Ottoman control of large parts of Central Europe

1708 - Charles XII of Sweden stops his march to conquer Moscow outside Smolensk, marking the turning point in the Great Northern War. The army is defeated nine months later in the Battle of Poltava, and the Swedish empire is no longer a major power.

1709 - Battle at Malplaquet during War of the Spanish Succession: Engl/Aust/Dutch Great Alliance beat France

1714 - French & Spanish troops under Duke of Berwick occupy Barcelona

1741 - Queen Maria Theresa addresses Hungarian Parliament

1758 - Battle of Saint Cast France repels British invasion during the Seven Year's War.

1773 - Benjamin Franklin writes "There never was a good war or bad peace"

1777 - Battle of Brandywine, Pa; Americans lose to British

1786 - Annapolis Convention to determine interstate

commerce, and to revise Articles of Confederation.
1789 - Alexander Hamilton appointed 1st Secretary of Treasury
1792 - The Hope Diamond is stolen with other crown jewels when six men break into house used to store the jewels.
1802 - France annexes the Kingdom of Piedmont.
1814 - Begun on the 6th Sept, Americans defeat British at Battle of Plattsburgh during War of 1812
1814 - Battle of Lake Champlain, NY; American Navy defeats British
1831 - Charles Darwin meets with capt Fitzroy at Plymouth
1852 - Olympia Columbian is 1st newspaper published north of Columbia R
1853 - 1st electric telegraph used (Merchant's Exchange to Pt Lobos)
1857 - Mountain Meadows Massacre, Mormons dressed as Indians murder 120 colonists in Utah
1881 - Triple landslides bury Elm Switz
1883 - James Cutler patents postal mail chute
1885 - Moses Hopkins, named minister to Liberia
1888 - Death of the Argentine politician Domingo Sarmiento, after whom the Latin American Teacher's Day is chosen.
1891 - The Jewish Colonization Association is established by Baron Maurice de Hirsch.
1893 - Bronx Gas & Electric Company opens on Frisby & Tremont Ave
1893 - Shaku Soen is 1st Zen teacher to visit the West (Chicago)
1893 - First conference of the World Parliament of Religions is held.
1895 - FA Cup stolen in Birmingham
1897 - After months of pursuit, generals of Menelik II of Ethiopia capture Gaki Sherocho, the last king of Kaffa,

bringing an end to that ancient kingdom.
1900 - President Paul Kruger crosses border into Mozambique
1906 - Mahatma Gandhi coins the term "Satyagraha" to characterize the Non-Violence movement in South Africa.
1909 - Max Wolf rediscovers Halley's comet
1910 - 1st commercially successful electric bus line opens (Hollywood)
1914 - Australia invades New Britain, defeating a German contingent there.
1916 - German troops conquer Kavalla Greece
1916 - The Quebec Bridge's central span collapses during reconstruction, killing 11 men, following initially total collapsed on August 29, 1907.
1919 - US marines invade Honduras
1921 - Nahalal, the first moshav in Israel, is settled.
1922 - British mandate of Palestine begins
1922 - Yankees play their farewell home game in Polo Grounds win doubleheader
1922 - The Treaty of Kars is ratified in Yerevan, Armenia.
1922 - One of the Herald Sun of Melbourne, Australia's predecessor papers The Sun News-Pictorial is founded.
1926 - Aloha Tower dedicated in Honolulu
1926 - Spain leaves League of Nation due to Germany joining
1930 - Stromboli volcano (Sicily) throws 2-ton basaltic rocks 2 miles
1931 - Salvatore Maranzano is murdered by Charles Luciano's hitmen.
1932 - Franciszek Żwirko and Stanisław Wigura, Polish Challenge 1932 winners, are killed in a plane crash when their RWD 6 crashes into the ground during a storm.
1936 - FDR dedicates Boulder Dam, now known as Hoover Dam

1939 - Battle of Kutno-pocket: Germans advance to Warsaw
1939 - British submarine Triton torpedoes British submarine Oxley
1939 - Iraq & Saudi Arabia declare war on nazi-Germany
1940 - Buckingham Palace damaged by German bombs
1940 - 54th US Postmaster General: Frank C Walker of Pa takes office
1940 - Anton Mussert establishes Dutch SS
1940 - Hitler begins operation Seelöwe (Sealion - aborted invasion England)
1940 - George Stibitz pioneers the first remote operation of a computer.
1941 - FDR orders any Axis ship found in American waters be shot at on sight
1941 - Charles Lindbergh, charges "British, Jewish & Roosevelt administration" are trying to get US into WW II, and speaks out against the Federal Reserve system.
1941 - World War II: The U.S. Navy is ordered to attack German U-boats.
1941 - Construction of the Pentagon begins (completed 15 Jan 1943)
1942 - Transport nr 31 departs with French Jews to Nazi-Germany
1943 - Allied arm forces conquerors Salerno
1943 - Jewish ghettos of Minsk & Lida Belorussia liquidated
1943 - Last German Q/pirate ship sinks near Easter Island
1943 - US & Australian troops join in Salamaua, New Guinea
1944 - FDR & Churchill meet in Canada at 2nd Quebec Conference
1944 - US 5th Armored Division is 1st allied force to enter nazi-Germany
1945 - Physician Willem J Kolff performs the first successful kidney dialysis using his artificial kidney machine, the

Netherlands

1946 - 1st mobile long-distance car-to-car telephone conversation

1952 - West German Chancellor Adenauer signs a reparation pact for Jews

1953 - *KSBW TV channel 8 in Salinas-Monterey, CA (NBC) begins broadcasting*

1953 - *WEHT TV channel 25 in Evansville, IN (CBS) begins broadcasting*

1954 - *1st Miss America TV broadcast*

1954 - 20.4 cm rainfall at Brunswick, Maine (state record)

1954 - *KXJB TV channel 4 in Valley City (Fargo) (NBC) begins broadcasting*

1954 - *Lee Meriwether (California), 19, crowned 27th Miss America 1955 (1st on TV)*

1955 - *KTVT TV channel 11 in Fort Worth-Dallas, TX (IND) begins broadcasting*

1955 - *Dedication of the first Temple of The Church of Jesus Christ of Latter-day Saints in Europe, the Bern Switzerland Temple.*

1958 - *Great Britain performs atmospheric nuclear test at Christmas Island*

1959 - "Duke" Ellington wins Springarn Medal for his musical achievements

1959 - Congress passes a bill authorizing food stamps for poor Americans

1960 - 17th Olympic games close in Rome Italy

1960 - The Young Americans for Freedom, meeting at home of William F. Buckley, Jr., promulgate the Sharon Statement.

1961 - Foundation of the World Wildlife Fund.

1962 - *KVCR TV channel 24 in San Bernardino, CA (PBS) begins broadcasting*

1963 - Typhoon Gloria strikes Taiwan killing 330, with $17.5

million damage

1965 - The 1st Cavalry Division of the United States Army arrives in Vietnam.

1966 - *France performs nuclear test at Muruora Island*

1967 - French president De Gaulle visits Poland

1967 - Indian/Chinese border fights

1967* - *US Surveyor 5 makes 1st chemical analysis of lunar material

1967 - *WSRE TV channel 23 in Pensacola, FL (PBS) begins broadcasting*

1967 - *WUNE TV channel 17 in Linville, NC (PBS) begins broadcasting*

1967 - *WUNF TV channel 33 in Asheville, NC (PBS) begins broadcasting*

1967 - *WUNG TV channel 58 in Concord, NC (PBS) begins broadcasting*

1968 - Air France Flight 1611 crashes off Nice, France, killing 89 passengers and 6 crew.

1969 - *USSR performs nuclear test at Eastern Kazakh/Semipalitinsk USSR*

1970 - **88 of the hostages from the Dawson's Field hijackings are released. The remaining hostages, mostly Jews and Israeli citizens, are held until September 25**

1971 - **Nakita Kruschev dies at 77**

1971 - **Egypt adopts its constitution**

1972 - **Munich Olympic games end after 11 are murdered**

1972 - BART begins service with a 26-mi (42-km) line from Oakland to Fremont

1973 - *Chile's President Salvador Allende deposed in a military coup*

1980 - *Chile adopts its constitution*

1981 - 2nd government of Van Agt forms in Netherlands

1983 - *USSR performs nuclear test at Eastern*

Kazakh/Semipalitinsk USSR

1985 - Intl Cometary Explorer (ISEE 3) passes Giacobini-Zinner by 7900 km

1986 - Pres Mubarak receives Israeli premier Peres

1986 - US performs nuclear test at Nevada Test Site

1986 - Dow Jones Industrial Avg suffered biggest 1-day decline ever, plummeting 86.61 points to 1,792.89. 237.57 million shares traded

1987 - Shoot out at Jean-Bertrand Aristides' church in Haiti, 12 die

1988 - 1/3 of population argues for Estonia autonomy

1989 - Drexel formally pleads guilty to security fraud

1989 - KSO-AM in Des Moines Iowa changes call letters to KGGO **1990 - George Bush HW Bush delivers his New World Order speech before Congress**

1991 - 14 die in a Continental Express commuter plane crash near Houston

1991 - Air crash at Djeddah, Saudi-Arabia, 263 die

1995 - Soyuz TM-22, lands

1996 - Union Pacific Railroad purchases Southern Pacific Railroad.

1997 - After a nationwide referendum, Scotland votes to establish a devolved parliament, within the United Kingdom.

1998 - Independent counsel Kenneth Starr sends a report to the U.S. Congress accusing President Bill Clinton of 11 possible impeachable offenses.

1998 - Opening ceremony for the 1998 Commonwealth Games in Kuala Lumpur, Malaysia. Malaysia is the first Asian country to host the games.

2000- Activists protest against the World Economic Forum meeting in Melbourne, Australia.

2001 - Two passenger planes hijacked by terrorists crashes into New York's World Trade Towers causing the collapse of

both & death of 2,752 people
2001 - Terrorists hijack a passenger plane and crash it into the Pentagon causing the death of 125 people
2001 - Attempt by passengers and crew of United Airlines Flight 93 to retake control of their hijacked plane from terrorists causes plane to crash in Pennsylvania field killing all 64 people onboard
2002 - Through extreme and coordinated effort, The Pentagon is rededicated after repairs are completed, exactly one year after the attack on the building.

911 Planes:

Flight #	*Reduced #*	*Crash Area*	*# of People*
?	*?*	*?*	*?*
11	*1+1=2*	*WTC*	*87*
93	*9+3=12 or 3*	*Shanksville*	*40*
175	*!+7+5=13 or 4*	*WTC*	*60*
77	*7+7=14 or 5*	*Pentagon*	*59*

What happened to plane number 1? This operation was designed based on Gematria and the significance of numbers. The first commercial flight (disputed) is said to have left Tampa, FL to Lexington, KY, with members of the Saudi Royal family in a Lear Jet 35, with a registration number of 505RP, or 5+5=10 or 1!

2003 - Swedish foreign minister Anna Lindh dies after being assaulted and fatally wounded on September 10.
2003 - The Cartagena Protocol on Biosafety enters into effect.
2004 - All passengers are killed when a helicopter crashes in the Aegean Sea. Passengers include Patriarch Peter VII of Alexandria and 16 others (including journalists and bishops

of the Greek Orthodox Church of Alexandria).

2005 - The State of Israel completes its unilateral disengagement from the Gaza Strip.

2007 *- Russia tests the largest conventional weapon ever, the Father of all bombs.*

2012 - 50 Al-Shabaab fighters are killed in conflict with the Somali National Army

2012 - The US consulate in Benghazi, Libya, is stormed, looted and burned down, killing five people, including the US ambassador

2012 - The US is warned by Moody's that its AAA credit rating is at risk if lawmakers fail to produce a long-term debt reduction plan

2012 - 29 people are killed and 11 injured after a bus runs off a mountain highway in Nepal

2012 - Japan nationalizes three of the disputed Senkaku Islands

2013 - 12 Alawite sect members are killed by rebel fighters in central Syria

2014 - South African athlete Oscar Pistorius is found not guilty of murdering his girlfriend Reeva Steenkamp (and is later found guilty of culpable homicide)

SEPTEMBER 21st ***261° tangent +6.31375151***
Autumnal Equinox

1192 - English king Richard I the Lion hearted, captured by by Leopold V, Duke of Austria

1217 - The Estonian tribal leader Lembitu of Lehola was killed in a battle against Teutonic Knights.

1348 - Jews in Zurich Switzerland are accused of poisoning wells

1435 - Treaty of Atrecht: Philip of Bourgondy vs French King Charles II

1451 - Cardinal Nicholas of Cusa orders Jews of Holland to wear a badge
1589 - Battle at Arques: French King Henri IV beats Catholic League
1591 - French bishops recognize Henri IV as king of France
1621 - King James I of England gives Sir Alexander Sterling royal charter for colonisation of Nova Scotia
1648 - Battle at Pilawce: Bohdan Chmielricki's beats John Casimir
1676 - Benedetto Odescalchi elected as Pope Innocent XI
1677 - John & Nicolaas van der Heyden patents fire extinguisher
1745 - Battle of Preston Pans: Bonnie Prince Charles beats English army
1746 - French expeditionary army occupies Labourdonnais & Dupleix Madras
1765 - Antoine de Beauterne announces he had killed the Beast of Gévaudan, but was later proved wrong by more attacks.
1776 - 5 days after British take NY, a ¼ of city burns down
1776 - Nathan Hale, spied on British for American rebels, arrested
1780 - Benedict Arnold gives British Major Andre plans to West Point
1784 - 1st daily newspaper in America (Pennsylvania Packet & General Advertiser)
1792 - Collot D'Herbois, first man to propose to abolish the monarchy in France at National Convention
1815 - King Willem I takes oath in Brussels
1823 - Moroni 1st appears to Joseph Smith, according to Smith
1827 - According to Joseph Smith Jr., the angel Moroni gave him a record of gold plates, one-third of which Joseph

translated into The Book of Mormon

1837 - Charles Tiffany founded his jewelry & china stores

1860 - In the Second Opium War, an Anglo-French force defeats Chinese troops at the Battle of Baliqiao.

1863 - Union forces retreat to Chattanooga after defeat at Chickamauga

1872 - John Henry Conyers of SC becomes 1st black student at Annapolis

1883 - 1st direct US-Brazil telegraph connection

1885 - Dutch demonstrate for general voting right

1893 - Frank Duryea drives 1st US made gas propelled vehicle (car)

1895 - 1st auto manufacturer opens-Duryea Motor Wagon Company

1896 - General Kitchener's army occupies Dongola Sudan

1897 - NY Sun runs famous "Yes, Virginia there is a Santa Claus," editorial

1898 - Empress Dowager Cixi seizes power and ends the Hundred Days' Reform in China, imprisoning the Guangxu Emperor.

1903 - 1st cowboy film "Kit Carson" premieres in US

1904 - The general strike called by the Socialist Party that spread throughout Italy ends today

1905 - Atlanta Life Insurance Co forms

1911 - Canadian Prime Minister Wilfrid Laurier loses the election to Robert Bordon of the Conservative Party

1913 - 1st aerobatic maneuver, sustained inverted flight, performed in France

1913 - Turkey & Bulgaria sign peace treaty in Constantinople

1915 - CH Chubb buys Stonehenge for £6,600

1915 - Emanuel Querido ("Kerido") begins publishing Querido

1919 - 33rd US Womens Tennis: Hazel H Wightman beats M Zinderstein

1921 - Pope Benedictus XV donates 1 million lire to feed Russians

1921 - Oppau explosion, a storage silo at a BASF fertilizer producing plant explods in Oppau, Germany, 500—600 killed.

1922 - Pres Warren G Harding signs a joint resolution of approval to establish a Jewish homeland in Palestine

1928 - "My Weekly Reader" magazine made its debut

1930 - Johann Ostermeyer patents flashbulb

1931 - Britain abandons gold standard; pound devalues 20%

1933 - Trial against Marinus der Lubbe opens

1934 - Typhoon strikes Honshu Island Japan, kills 4,000

1936 - Spanish fascist junta names Franco to generalissimo/supreme commander

1937 - J. R. R. Tolkien's 'The Hobbit' is published

1938 - Hurricane (183 MPH winds) in New England kills 700

1938 - Winston Churchill condemns Hitler's annexation of Czechoslovakia

1938 - The Great Hurricane of 1938 makes landfall on Long Island in New York. The death toll is estimated at 500-700 people.

1939 - Reinhard Heydrich meets in Berlin to discuss final solution of Jews

1941 - US launches its 1st Liberty-ship, "Patrick Henry"

1942 - 116 hostages executed by Nazis in Paris

1942 - Transport nr 35 departs with French Jews to nazi-Germany

1943 - Arundel (Solomon Island) in US hands

1943 - Lynch Triangle (Square) in Bronx named

1943 - Russian 13th/61st Army reconquer Chyernigov

1943 - Soviet forces reach Dnjepr

1944 - Last British paratroopers at bridge of Arnhem surrenders

1949 - Chinese Communist leaders proclaim People's

Republic of China
1949 - Federal Republic of [West] Germany created under 3-power occupation
1950 - George Marshall sworn in as the 3rd Secretary of Defense of United States.
1951 - Emil Zatopek runs 15,000m in record 44 min, 54.6 sec
1953 - Allied forces form West Germany
***1953** - KRDO TV channel 13 in Colorado Spgs-Pueblo, CO (ABC) 1st broadcast*
1954 - 1st nuclear submarine, USS Nautilus, commissioned under Commander Eugene P. Wilkinson
1954 - Kleffens appointed chairman of General Meeting UN
1955 - Last allied occupying troops leave Austria
***1955** - USSR performs nuclear test*
1957 - German sail training ship Pamir sails Atlantic Ocean
1957 - Olav V becomes king of Norway
1957 - Pote Sarasin forms government in Thailand
1958 - 1st airplane flight exceeding 1200 hours, lands, Dallas Tx
1958 - US performs nuclear test at Nevada Test Site
1959 - 600 Indian Dutch emigrate to US
1961 - Maiden flight of the CH-47 Chinook transportation helicopter.
1964 - Malta becomes independent from the United Kingdom.
1964 - The North American XB-70 Valkyrie, the world's first Mach 3 bomber, made its maiden flight from Palmdale, California.
1965 - Singapore admitted as part of the United Nations.
1966 - 5inches of rain falls on NYC
1966 - Jimmy Hendrix changes spelling of his name to Jimi
1967 - US performs nuclear test at Nevada Test Site
1970 - "Monday Night Football" premieres on ABC - Browns

31, Jets 21

1970 - *KAPP TV channel 35 in Yakima, WA (ABC) begins broadcasting*

1970 - Luna 16 leaves Moon

1970 - New York Times starts first modern op-ed page.

1972 - Marcos declares martial law in Philippines

1972 - *USSR performs underground nuclear test*

1974 - US Mariner 10 makes 2nd fly-by of Mercury

1976 - Orlando Letelier is assassinated in Washington, D.C. He was a member of the Chilean socialist government of Salvador Allende, overthrown in 1973 by Augusto Pinochet.

1977 - US minister of Foreign affairs Cyrus Vance dismissed

1978 - The Provisional Irish Republican Army explode bombs at the RAF airfield near Eglinton, County Londonderry; the terminal building, two aircraft hangars and four planes aree destroyed

1979 - Two RAF Hawker Siddeley Harrier jump-jets from RAF Wittering collide over the UK. Both pilots ejected safely. One of the jets broke up in midair and fell harmlessly into a field but the other dropped onto the centre of Wisbech in Cambridgeshire, destroying two houses and a bungalow. Several people were injured in the accident and three people were killed.

1979 - Stephen Jay Gould and Richard Lewontin publish their influential paper (cited more than 4,000 times) "The Spandrels of San Marco and the panglossian paradigm", introducing the term and idea of "spandrels" into evolutionary biology

1981 - Belize (British Honduras) gains independence from UK

1981 - Sandra Day O'Conner becomes 1st female Supreme Court Justice

1981 - Enterprise Radio (all sports) goes off the air

1982 - STS-5 vehicle moves to launch pad

1983 - 11 killed in anti Marcos demonstrations in Manila

1983 - US performs nuclear test at Nevada Test Site

1984 - NASA launches Galaxy-C

1987 - 3 Belgian minesweepers depart to Persians Gulf

1989 - Poland's Sejm (National Assembly) approves prime minister Mazowiecki

1990 - Reports that US refinery problems will lead to a loss in capacity and aggressive remarks by Saddam Hussein send crude prices to new highs

1991 - Armenia votes on whether to remain in Soviet Union

1991 - USA Basketball announces "Dream Team" for the 1992 Olympics

1993 - Ukraine government of Kutshma resigns

1995 - The Hindu milk miracle occurs, in which statues of the Hindu God Ganesh began drinking milk when spoonfuls were placed near their mouths.

1999 - Chi-Chi earthquake occurs in central Taiwan, leaving about 2,400 people dead.

2001 - Deep Space 1 flies within 2,200 km of Comet Borrelly.

2001 - University of Roorkee, becomes India's 7th Indian Institute of Technology, rechristened as IIT Roorkee

2001 - AZF chemical plant explodes in Toulouse, France, killing 29 people

2003 - Galileo mission terminated by sending the probe into Jupiter's atmosphere, where it is crushed by the pressure at the lower altitudes.

2004 - The Communist Party of India (Marxist-Leninist) People's War and the Maoist Communist Centre of India merge to form the Communist Party of India (Maoist).

2004 - Construction of the Burj Dubai starts.

2008 - Goldman Sachs and Morgan Stanley, the two last remaining independent investment banks on Wall Street, become bank holding companies as a result of the subprime

mortgage crisis.

2008 - President Thabo Mbeki of South Africa resigns from office, effective September 25.

2008 - Prime Minister Ehud Olmert of Israel formally resigns from office, effective as soon as his successor Tzipi Livni has successfully assembled a new government.

2009 - Macedonia officially enters the recession after experiencing a drop in GDP of 0.9% in the first quarter, followed by 1.4% in the second quarter

2012 - Japanese Prime Minister, Yoshihiko Noda, wins a leadership ballot

2013 - 6 people are killed in military and insurgent conflict in Zamboanga City, Philippines

2013 - 100 people are killed in a series of attacks across Iraq

2013 - 59 people are killed and 175 wounded in a shopping mall gun battle in Nairobi, Kenya

DECEMBER 21st ***351° - tangent -0.1583844***
Winter Solstice

1163 - Hurricane hits villages in Holland/Friesland, causing floods

1561 - Antoine Perrenot de Granvelle, Archbishop of Mechelen, made Cardinal

1582 - Flanders adopts Gregorian calendar, tomorrow is Jan 1 1583

1598 - Battle of Curalaba: The revolting Mapuche, led by cacique Pelentaru, inflict a major defeat on Spanish troops in southern Chile; all Spanish cities south of the Biobio river are eventually taken by the Mapuches, and all conquest of Mapuche territories by Europeans practically ceases, until the 1870s "Pacification of Araucania".

1620 - 103 Mayflower pilgrims land at Plymouth Rock

1650 - Johan de Witt installed as Dutch pension advisor of

Dordrecht

1784 - John Jay becomes 1st US Secretary of State (foreign affairs)

1788 - Hue Tay Son becomes emperor Quang Trung of Vietnam

1829 - 1st stone arch railroad bridge in US dedicated, Baltimore

1835 - HMS Beagle sails into Bay of Islands (New Zealand)

1844 - The Rochdale Pioneers commence business at their cooperative in Rochdale, England, starting the Cooperative movement.

1864 - General Sherman conquers Savannah, Georgia

1866 - Cheyennes, Arapho's, Sioux, Fetterman Massacre

1890 - Pim Mulier 1st & only trip to "Alvesteddetocht"

1894 - Mackenzie Bowell becomes the 5th Prime Minister of Canada

1898 - Scientists Pierre and Marie Curie discover radium

1906 - British Parliament pass two important pieces of social legislation: The Trades Disputes Bill, legalizing peaceful picketing, and The Workingmen's Compensation Act, broadening employers' liability for accidents

1907 - Dutch government of De Master falls due to war budget

1909 - U of Coopenhagen rejects Cook's claim that he was 1st to North Pole

1910 - Explosion in coal mine in Hulton England, 344 mine workers dies

1912 - Denmark, Norway & Sweden declare neutrality in Comende war

1914 - 2,800 African miners strike at the Van Rhyn Deep mines in a bid to redress some of their grievances

1915 - 10.17" (25.83 cm) of rainfall, Glenora, Oregon (state record)

1919 - J. Edgar Hoover deports anarchists/feminist Emma Goldman to Russia
1921 - Supreme Court rules labor injunctions & picketing unconstitutional
1923 - Nepal changes from British protectorate to independent nation
1925 - Eisenstein's movie Battleship Potemkin premieres in Moscow
1929 - 1st group hospital insurance plan offered (Dallas Tx)
1933 - Dried human blood serum 1st prepared, Univ of Pennsylvania
1933 - Newfoundland reverts to being a crown colony of Great Britain
1937 - 1st feature-length color & sound cartoon premieres (Snow White)
1937 - The first full-length cel animated feature film and the earliest in the Walt Disney Animated Classics series, "Snow White and the Seven Dwarfs", premieres at the Carthay Circle Theatre
1939 - Hitler named Adolf Eichmann leader of "Referat IV B"
1941 - German submarine U-567 sinks
1941 - Last NFL drop kick for an extra point (Ray McLean, Chicago Bears)
1942 - US Supreme court declares Nevada separation legal
1946 - Earthquake in South Japan, kills 1,086
1948 - State of Eire (formerly Irish Free State) declares its independence
1949 - Dutch 1st Chamber accept soeveregnty of Indonesia
1952 - Broadway Tunnel opens in SF
1952 - WSBT TV channel 22 in South Bend, IN (CBS) begins broadcasting
1953 - KOMU TV channel 8 in Columbia, MO (NBC/PBS) begins broadcasting

1957 - Indonesia proclaims end to state of war
1958 - Charles de Gaulle wins 7 year term as 1st pres of 5th Rep of France
1959 - 10th largest snowfall in NYC history (13.7")
1959 - Citizens of Deerfield Ill block building of interracial housing
1961 - JFK & British PM MacMillan meet in Bermuda
1961 - Gangster Joe Gallo is sentenced to 7 to 14 years in state prison for extortion
1962 - Angolin leaves Comecon
1962 - US & Cuba accord, releases bay of pigs captive
1966 - USSR launches Luna 13; soft-landed in Oceanus Procellarum
1968 - Apollo 8 (Borman, Lovell & Anders) 1st manned Moon voyage
1970 - WUTV TV channel 29 in Buffalo, NY (IND) begins broadcasting
1971 - UN Security Council chooses Kurt Waldheim as 4th secretary General
1971 - A publican is killed as he tried to remove a bomb from his pub, Northern Ireland
1972 - Soviet Union signs a separate peace with East Germany
1973 - Israel, Egypt, Syria, Jordan, US & USSR meet in Geneva to discuss the Arab-Israeli conflict
1975 - Madagascar adopts constitution
1976 - Patricia R Harris named secretary of HUD
1976 - UN General Assembly passes a resolution declaring 1979-Year of Child
1976 - US performs nuclear test at Nevada Test Site
1978 - Police in Des Plaines Ill, arrest John Wayne Gacy for murder
1979 - USSR performs nuclear test at Eastern

Kazakh/Semipalitinsk USSR

1979 - New constitution for Zimbabwe agreed

1984 - USSR launches Vega 2 for fly-by of Halley's Comet

1985 - ARCO Anchorage runs aground near Port Angeles, WA

1987 - 3 white NY teens convicted of manslaughter in death of a black man

1987 - Soyuz TM-4 launches 3 cosmonauts to space station Mir

1987 - Vladimir Titov & Musa Manarov launched

1988 - Drexel agrees guilt to security felonies, pays a $650 million fine

1988 - Vladimir Titov & Musa Manarov return to earth (a year) with Chretien

1988 - Lockerbie disaster: Pan Am Flight 103 destroyed mid air by a terrorist bomb killing all 258 on board over Scotland

1989 - VP Quayle sends out 30,000 Xmas cards with word beacon spelled beakon

1989 - New Zealand is the first country to set a formal inflation target (0-2%) for how much prices should rise each year. Similar targets are subsequently adopted by most developed countries.

1991 - 95 share in Madrid Spain $1.3 billion lottery (#47996)

1991 - El Sayid Nosair acquitted of killing Meir Kahane

1991 - Soviet Union formally dissolves as 11 of 12 republics sign treaty forming Commonwealth of Independent States

1992 - Dutch DC-10 in fire at landing on Faro Portugal, 56 die

1994 - Bomb goes off on #4 train on Fulton Street NYC

1995 - The city of Bethlehem passes from Israeli to Palestinian control.

1999- The Spanish Civil Guard intercepts a van loaded with 950 kg of explosives that ETA intended to use to blow up Torre Picasso in Madrid.

2007 - The Schengen Agreement area increases to include 9

European Union member states; Czech Republic, Estonia, Hungary, Latvia, Lithuania, Malta, Poland, Slovakia, and Slovenia on land and sea borders.
2012 - 39 people are killed in violent clashes in Kenya
2012 - The Mesoamerican Long Count calendar reaches the date 13.0.0.0.0

DECEMBER 25TH ***December 25 - 355° - tangent -0.0874887***

(Ro) Dies Natalis Sol Invictus (Birthday of the Unconquered Sun), and Brumalia. (Ca/LCa) The Nativity of Christ is held by the Catholic faithful. (EO) The Nativity of Christ is recognized. (US) Christmas held in the United States.

1 - 1st Christmas, according to calendar-maker Dionysus Exiguus
274 - Roman Emperor Aurelian dedicates a temple to Sol Invictus on the supposed day of the winter solstice and day of rebirth of the Sun.
337 - Earliest possible date that Christmas was celebrated on Dec 25th
352 - 1st definite date Christmas was celebrated on Dec 25th
390 - Roman emperor Theodosius admits debt on mass murder in Thessalonica
498 - French King Clovis baptises himself
597 - England adopts Julian calendar
604 - Battle at Etampes (Stampae): Burgundy beat Neustriers
800 - Pope Leo III crowns Charles the Great (Charlemagne), Roman emperor
875 - Charles the Bare crowned emperor of Rome
967 - John XIII crowned Otto II the Red German compassionate emperor
969 - Johannes I Tzimisces crowned emperor of Byzantium

979 - Rotardus appointed as bishop of the kingdom
999 - Heribertus becomes bishop of Cologne
1000 - Monarch Istvan crowned king of Hungary
1046 - Pope Clemens VI crowns Henry III as Holy Roman Emperor
1048 - Parliament of Worms: Emperor Henry III names his cousin count Bruno van Egisheim/Dagsburg as Pope Leo IX
1066 - Duke William of Normandy ('William the Conqueror) crowned king of England
1100 - Boudouin I of Boulogne crowned king of Jerusalem
1101 - Henry I of Limburg becomes duke of Neth-Lutherans
1121 - Norbertus van Xanten finds order of the Norbertijnen
1130 - Anti-pope Anacletus II crowns Roger II the Norman king of Sicily
1223 - St Francis of Assisi assembles 1st Nativity scene (Greccio, Italy)
1261 - John IV Lascaris of the restored Eastern Roman Empire is deposed and blinded by orders of his co-ruler Michael VIII Palaeologus.
1492 - Columbus' ship Santa Maria runs aground and sinks on Hispaniola
1522 - Turkish troops occupy Rhodos
1553 - Battle of Tucapel: Mapuche rebels under Lautaro defeats the Spanish conquistadors and exetutes the governor of Chile Pedro de Valdivia.
1582 - Zealand/Brabant adopts Gregorian calendar, yesterday was Dec 14th
1599 - The city of Natal, Brazil, is founded.
1613 - Johan Sigismund of Brandenburg becomes protestant
1621 - Gov William Bradford of Plymouth Colony (now in Massachusetts) forbids game playing on Christmas
1640 - Pierre de Fermat writes to Marin Mersenne about Fermat's church thesis

1641 - Emperor Ferdinand III makes appointments with Sweden & France

1643 - Christmas Island founded and named by Captain William Mynors of the East India Ship Company vessel, the Royal Mary.

1651 - Massachusetts General Court ordered a five shilling fine for "observing any such day as Christmas"

1683 - English Whig leader Duke of Monmouth flees to Holland

1688 - British king James II lands in Ambleteuse, France

1688 - Lord Delamere sides with King James II

1717 - Floods ravage Dutch coast provinces, thousands killed

1741 - Astronomer Anders Celsius introduces Centigrade temperature scale

1745 - Prussia and Austria sign Treaty of Dresden giving much of Silesia to the Prussians

1758 - Return of Halley's comet 1st sighted by Johann Georg Palitzsch

1760 - Jupiter Hammon, an African American slave, composes poetry broadside "An Evening Thought" (1st poetry published by an Amercian American 1761)

1769 - First Christian service in New Zealand; Mass is celebrated in Doubtless Bay by Father Paul-Antoine Léonard de Villefeix of the de Surville expedition

1775 - Pope Pius VI encyclical on the problems of the pontificate

1776 - Washington crosses Delaware & surprises & defeats 1,400 Hessians

1809 - Physician Ephraim McDowell performs the first abdominal surgery in the U.S, an ovariotomy to remove a 22 lb ovarian tumor

1814 - Rev Samuel Marsden of the Church Missionary Society holds the first Christian service in New Zealand on land, at

Rangihoua

1818 - 1st known Christmas carol ("Silent Night, Holy Night" - "Stille Nacht, Heilige Nacht) sung (Austria)

1818 - Handel's Messiah, premieres in the US in Boston

1831 - Louisiana & Arkansas are first states to observe Christmas as holiday

1832 - Charles Darwin celebrates Christmas in St Martin at Cape Receiver

1833 - Charles Darwin celebrates Christmas in Port Desire, Patagonia

1834 - Charles Darwin celebrates Christmas on Beagle at Tres Montes, Chile

1835 - Charles Darwin celebrates Christmas in Pahia, New Zealand

1837 - Battle of Okeechobee-US forces defeat Seminole Indians

1848 - New Haven Railroad opens

1868 - Despite bitter opposition, President A Johnson grants unconditional pardon to all persons involved in Southern rebellion (Civil War)

1896 - "Stars & Stripes Forever" written by John Philip Sousa

1899 - Farmers send/guide belegerd Ladysmith Kerstpudding/desire in grenade

1900 - Arthur Schnitzler's "Leutnant Gustl" forbidden in Germany

1901 - Battle at Tweefontein Orange-Free state: Boers surprise attack Brits

1901 - At Tweefontein, a British force is camped on a hill slope; at 2am it is attacked by Boers coming down from the other side

1902 - Pope Leo XIII, at his annual Christmas reception, endorses the Christian Democratic movement now emerging in Europe as an attempt to offer an alternative to more

radical movements
1914 - Legendary "Christmas Truce" takes place on the battlefields of WWI between British and German troops. Instead of fighting, soldiers exchange gifts and play football.
1922 -Dec 26] Lenin dictates his "Political testament"
1926 - Prince-regent Hirohito becomes Emperor of Japan (1926-1989) upon the death of his father Yoshihito
1932 - 7.6 magnitude earthquake ravages Qansu, China, kills 275
1932 - During King George V's Christmas dinner speech his chair collapes
1933 - Belgian Working people's party accept Henry de Mans Plan of Labor
1936 - Belgian bishops condemn fascism & communism
1939 - Montgomery Ward introduces Rudolph the 9th reindeer
1941 - Japan announces surrender of British-Canadian garrison at Hong Kong
1941 - Japanese aircraft carriers Akagi/Kagu back in Kure, Japan
1942 - Admiral Dalans murderer of Bosinier de la Chapelle, sentenced to death
1942 - British Col S W Bailey reaches Mihailovics headquarter
1942 - Russian artillery/tank battle on German armies at Stalingrad
1946 - Constitution accepted in Taiwan
1947 - Taiwan passes Human Rights laws (Day of Earth Law)
1947 - The Constitution of the Republic of China comes into effect.
1950 - Coronation Stone, taken from Scone in Scotland by Edward I in 1296, stolen from Westminster Abbey & smuggled back to Scotland
1953- Avalanche of lava kills 150 (Ruapehu volcano, New

Zealand)
1954 - WSFA TV channel 12 in Montgomery, AL (NBC) begins broadcasting
1955 - Pope Pius XII encyclical on sacred music & popular music
1957 - Ed Gein found insane of murder
1959 - A synagogue in Cologne Germany desecrated with swatstikas
1959 - Sony brings transistor TV 8-301 to the market
1962 - USSR performs nuclear test at Novaya Zemlya USSR
1963 - Walt Disney's "The Sword in the Stone" is released
1965 - The Yemeni Nasserite Unionist People's Organisation is founded in Taiz
1968 - Frank Borman's Christmas reading while orbiting Moon
1968 - 42 Dalits are burned alive in Kilavenmani village, Tamil Nadu, India, a retaliation for a campaign for higher wages by Dalit labourers.
1969 - 5 Israeli gunboats escape from Cherbourg harbor
1971 - Operation PUSH (People United to Save Humanity) form Jesse Jackson
1973 - The ARPANET crashes when a programming bug causes all ARPANET traffic to be routed through the server at Harvard University, causing the server to freeze.
1973 - Arab oil ministers cancel January 5 percent production cut; Saudi Arabian oil minister promises 10 percent OPEC production rise
1974 - Cyclone Tracy virtually destroys Darwin, Australia
1974 - Marshall Fields drives a vehicle through the gates of the White House, resulting in a four-hour standoff.
1976 - Egyptian SS Patria sinks in Red Sea, about 100 killed
1976 - Takeo Fukuda becomes Japanese premier
1977 - Israeli PM Menachem Begin meets Egyptian Pres

Sadat in Egypt

1979 - USSR airlifts invasionary army to Afghanistan

1983 - 1st live telecast of Christmas Parade at the EPCOT Centre, Disney World Florida

1987 - Lynette "Squeaky" Fromme, who escaped 2 days earlier, recaptured

1989 - Japanese scientist achieve -271.8°C, coldest temp ever recorded

***1989 - Show trial of Romanian Commuinst dicator Nicolae Ceaușescu and his wife Elena on charges of genocide and personal enrichment. The couple are found guilty and executed by* firing squad the same day.**

1990 - The first successful trial run of the system which would become the World Wide Web.

1991 - Mikhail Gorbachev formally resigns as President of USSR in a televised speech

2003 - The ill-fated Beagle 2 probe which was released from the Mars Express Spacecraft on December 19, disappears shortly before its scheduled landing.

2004 - Cassini orbiter releases Huygens probe which successfully landed on Saturn's moon Titan on January 14, 2005.

2009 - Umar Farouk Abdulmutallab unsuccessfully attempts a terrorist attack against the US while on board Northwest Airlines Flight 253

2012 - 27 people are killed after an Antonov An-72 plane crashes near Shymkent, Kazakhstan

2012 - 8 people are killed and thousands left homeless after two fires strike Manila, Philippines

Referenced from: www.historyorb.com

Conclusions

There should be no mistake made about the importance of Pi and Phi to the Golden Circle and the Golden Ellipse. Our annual trek around the solar system was setup with mathematic precision, and shows how the tangent and radian points factor into this journey.

The circle is dedicated to degrees, tangents, and radians, to establish position. Certain points, or sweet spots, are more favorable than others. The circle is divided into 4 equal portions, with two positive degree portions, and two negative degree portions. From 1 degree through 89 degree, we get positive tangent points. From 91 degree through 179 degree, we get negative tangent points. From 181 degree through 269 degree, we get positive tangent points. From 271 degree through 359 degree, we get negative tangent points. The 90, 180, 270, and 360 degree tangent points are considered zero points numerically, but have error based tangents. The main strength from these points come at the 72 degree and 252 degree points, as they mark the entrance into spring and fall. These are strong solar magnetic tangent points, as the axis of the planet aligns with the sun. We have also suffered major disasters in conjunction with these alignments in recent years, with the disappearance of Flight MH-370 on 3/8/2014 (68°), Fukushima earthquake, tsunami, and nuclear disaster on 3/11/2011 (71°), and the 9/11 tragedies on 9/11/2001 (251°). For 3 days or more on either side of a strong tangent point, the solar energy still reverberates. Remember that when someone understands something that you don't, then you are still in the dark!

We proceed from one tangent point to another during the course of our 24 hour day. At the 72 degree point of March 12th, we start the day at the +3.07768354 tangent point, and we reach the sweet spot of +3.1415 or Pi, during the path of the day. The opposite positive point of 252 degree on September 12th, is exactly 180 degrees from the 72 degree mark. This 252 degree point is also +3.07768354, and also reaches the Pi point of +3.1415 during the path of the day. This Pi feature also shows itself at the 107 degree point on April 17th (post Passover point!), with a tangent of -3.2708526, and also at the 287 degree of October 17th (Shemini Atzeret, Vijyadashami, Hajj), as we reach the negative Pi tangent point during the course of the day.

The Phi phenomenon exhibits itself at the 58 degree tangent point on 2/28, and at the 238 degree tangent point on 8/28, with a positive tangent of +1.60033453, and reaches Phi (1.618) during the journey of the day. The negative Phi tangent points are 121 degree of 5/1 (May Day, Beltane, Arbor Day), and 301 degree of 11/1 (All Saints, Samhain), and start the day at -1.6642795, and reach Phi during the day.

The product of Pi (3.1415) X Phi (1.1618) equals 5.0803. The positive Pi/Phi product points are 78 degree on 3/18 (Vernal Equinox 3/21), and the 258 degree on (Meditrinalia, Harvest Season), with a tangent of +4.70463011. The negative tangent points of the Pi/Phi product occur at 101 degree on 4/11 (Easter, Passover, Tax Day), and at the 281 degree on 10/11 (Meditrinalia/Harvest Season) with a tangent degree of -5.144554.

The four radian points on the circle are precise Pi points, as the 90 degree point (3/30) equals 1/2 π at 1.57079635, the 180 degree point (6/30) is the only point that equals π precisely at 3.14159265, the 270 degree point (9/30) equals Pi and a 1/2 of 4.71238898, and the 360 degree point (12/30) equals Pi doubled of 6.28318531.

Just as the zodiac signs overlap each month, so did the original monthly configuration. The Jewish (taken from the Babylonian system), Babylonian, Assyrian, and Levant calendars, were designed in the same overlapping format as the zodiac signs. This can be seen in the chart below:

30° Degrees	*Dates*	*Month*
15° *- 45°*	*Jan 15 to Feb 15*	
45° - 75°	*Feb 15 to March 15*	
75° - 105°	*March 15 to April 15*	*81° vernal equi*
105° *- 135°*	*April 15 to May 15*	
135° - 165°	*May 15 to June 15*	
165° - 195°	*June 15 to July 15*	*171°summer sol* *184° Aphelion*
195° *- 225°*	*July 15 to Aug 15*	
225° - 255°	*Aug 15 to Sept 15*	
255° - 285°	*Sept 15 to Oct 15*	*261° autumn equi*
285° *- 315°*	*Oct 15 to Nov 15*	
315° - 345°	*Nov 15 to Dec 15*	
345° - 15°	*Dec 15 to Jan 15*	*351° winter sol* *3° Perihelion*

We know that our proximity to the Sun does not account for the seasons, but the angle of the Earth relative to the axis. The degree of the Golden Ellipse of orbit signals the change to these seasons. The heavens have been squared, with the 15°, 105°, 195°, and 285° points being all important for this process. The vernal equinox occurs 24 degrees before the 105° point. The summer solstice occurs exactly 24 degrees before the 195° point. The autumnal equinox follows suit exactly 24 degrees before the 285° mark. The winter solstice occurs precisely 24 degrees before the 15° degree point. The heavens were originally setup for the 360 degrees or days of the circle, and the 365 day calendar was an adjustment after the cosmic disaster of 2344 BCE.

Pi and Phi, are all too important to the physiology of the human body, plant kingdom, engineering, science, celestial mechanics, and religion. Yes, religion! Our faith based systems have had their holi-days aligned with the ancient understanding of the workings of the solar system. While it is true that we have entirely too many holi-days on the calendar, it is by no accident that they have been aligned to these important tangent points. The clergy that took part in reconciling the calendars after these disasters, were very intelligent men, with a broad base of understanding. The issue with knowledge is, is it being used to encumber, or is it being used to liberate? There are few questions that religion has left. Do you worship the sun, do you worship the moon , or do you worship another heavenly body?

This lack of internal (esoteric) knowledge has left us with a modern Pagan calendar.

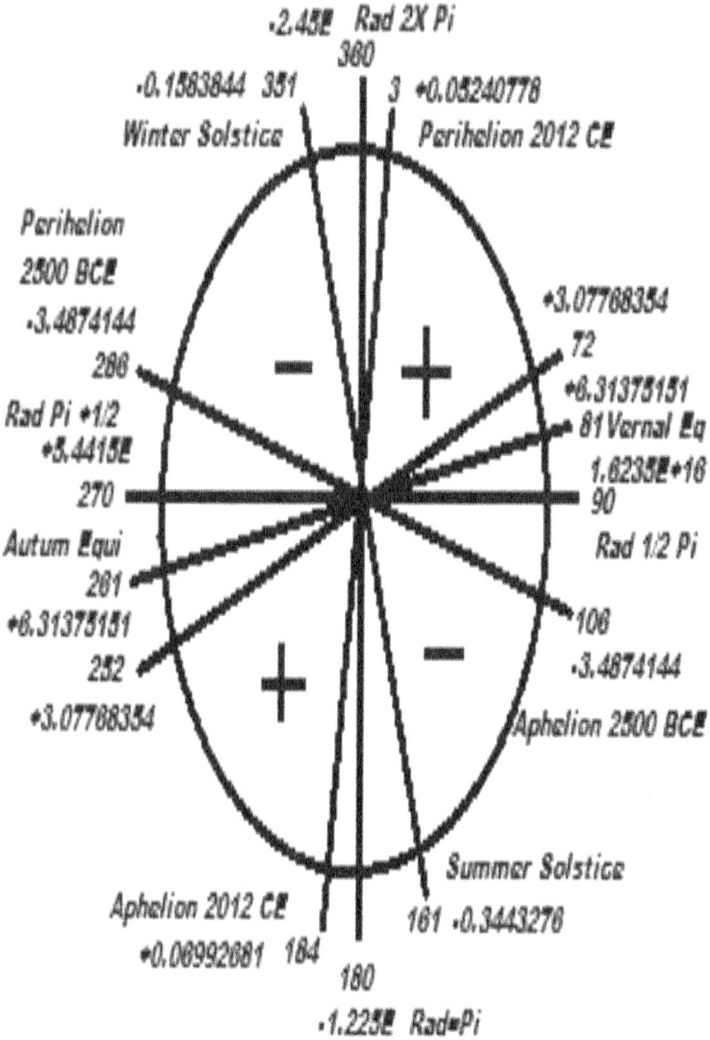

The Golden Circle and Ellipse

Key:

Ca - Catholic	*SR - Sacred*
Round	
LCa - Latin Catholic	*VC - Venus*
Calendar	
EO - Eastern Orthodox	*Ru - Russia*
Ch - Christian	*Ja - Japan*
Ro - Roman	
Hi - Hindu	
Bu - Buddhist	
P - Pagan	
W - Wiccan	
US - United States	
Lu - Lutheran	
Intl - International	
Is - Islamic	
J - Judaism	
We - Welsh	

360 VS 365 Day Calendars Side By Side

DEGREE 360	MONTH/DAY 360	JULIAN/365 DAY	Holiday or Solar Point
1	1-Jan	1-Jan	
2	2	2	
3	3	3	Perihelion Solemnity of Mary +0.05240778
4	4	4	
5	5	5	
6	6	6	
7	7	7	
8	8	8	
9	9	9	
10	10	10	
11	11	11	
12	12	12	
13	13	13	
14	14	14	
15	15	15	
16	16	16	
17	17	17	
18	18	18	
19	19	19	
20	20	20	
21	21	21	
22	22	22	

23	*23*	*23*	
24	*24*	*24*	
25	*25*	*25*	
26	*26*	*26*	
27	*27*	*27*	
28	*28*	*28*	
29	*29*	*29*	
30	*30*	*30*	
31	*1-Feb*	*31*	
32	*2*	*1-Feb*	*Ground Hog Day* *Imbolc* *+0.62486935*
33	*3*	*2*	
34	*4*	*3*	
35	*5*	*4*	
36	*6*	*5*	
37	*7*	*6*	
38	*8*	*7*	
39	*9*	*8*	
40	*10*	*9*	
41	*11*	*10*	
42	*12*	*11*	
43	*13*	*12*	
44	*14*	*13*	
45	*15*	*14*	*Lupercalia* *+1*
46	*16*	*15*	
47	*17*	*16*	
48	*18*	*17*	

49	*19*	*18*	
50	*20*	*19*	
51	*21*	*20*	
52	*22*	*21*	
53	*23*	*22*	
54	*24*	*23*	
55	*25*	*24*	
56	*26*	*25*	
57	*27*	*26*	
58	*28*	*27*	
59	*29*	*28*	
60	*30*	*1-Mar*	
61	*1-Mar*	*2*	*Roman New Years +1.80404776*
62	*2*	*3*	
63	*3*	*4*	
64	*4*	*5*	
65	*5*	*6*	
66	*6*	*7*	
67	*7*	*8*	
68	*8*	*9*	
69	*9*	*10*	
70	*10*	*11*	
71	*11*	*12*	
72	*12*	*13*	*Solar magnetic +3.0776*
73	*13*	*14*	
74	*14*	*15*	
75	*15*	*16*	*Passover/Conceptio n +3.73205081*

76	*16*	*17*	
77	*17*	*18*	
78	*18*	*19*	
79	*19*	*20*	
80	*20*	*21*	
81	*21*	*22*	*Vernal Equinox +6.3138*
82	*22*	*23*	
83	*23*	*24*	
84	*24*	*25*	
85	*25*	*26*	
86	*26*	*27*	
87	*27*	*28*	
88	*28*	*29*	
89	*29*	*30*	
90	*30*	*31*	*Radian = 1/2 Pi 1.57079635*
91	*1-Apr*	*1-Apr*	*Planting day -57.289962*
92	*2*	*2*	
93	*3*	*3*	
94	*4*	*4*	
95	*5*	*5*	
96	*6*	*6*	
97	*7*	*7*	
98	*8*	*8*	
99	*9*	*9*	
100	*10*	*10*	
101	*11*	*11*	
102	*12*	*12*	
103	*13*	*13*	

104	14	14	
105	15	15	Crucifixion -3.7320508
106	16	16	
107	17	17	
108	18	18	
109	19	19	
110	20	20	
111	21	21	
112	22	22	
113	23	23	
114	24	24	
115	25	25	
116	26	26	
117	27	27	
118	28	28	
119	29	29	
120	30	30	
121	1-May	1-May	May Day/Beltane -1.6642795
122	2	2	
123	3	3	
124	4	4	
125	5	5	
126	6	6	
127	7	7	
128	8	8	
129	9	9	
130	10	10	
131	11	11	

132	*12*	*12*	
133	*13*	*13*	
134	*14*	*14*	
135	*15*	*15*	
136	*16*	*16*	
137	*17*	*17*	
138	*18*	*18*	
139	*19*	*19*	
140	*20*	*20*	
141	*21*	*21*	
142	*22*	*22*	
143	*23*	*23*	
144	*24*	*24*	
145	*25*	*25*	
146	*26*	*26*	
147	*27*	*27*	
148	*28*	*28*	
149	*29*	*29*	
150	*30*	*30*	
151	*1-Jun*	*31*	
152	*2*	*1-Jun*	
153	*3*	*2*	
154	*4*	*3*	
155	*5*	*4*	
156	*6*	*5*	
157	*7*	*6*	
158	*8*	*7*	
159	*9*	*8*	
160	*10*	*9*	
161	*11*	*10*	

162	*12*	*11*	
163	*13*	*12*	
164	*14*	*13*	
165	*15*	*14*	
166	*16*	*15*	
167	*17*	*16*	
168	*18*	*17*	
169	*19*	*18*	
170	*20*	*19*	
171	*21*	*20*	*Summer Solstice -0.1583844*
172	*22*	*21*	
173	*23*	*22*	
174	*24*	*23*	
175	*25*	*24*	
176	*26*	*25*	
177	*27*	*26*	
178	*28*	*27*	
179	*29*	*28*	
180	*30*	*29*	*Radian = Pi 3.14159265*
181	*1-Jul*	*30*	
182	*2*	*1-Jul*	
183	*3*	*2*	
184	*4*	*3*	*Perihelion - 4th July +0.06992681*
185	*5*	*4*	
186	*6*	*5*	
187	*7*	*6*	
188	*8*	*7*	
189	*9*	*8*	

190	10	9	
191	11	10	
192	12	11	
193	13	12	
194	14	13	
195	15	14	17th Tammuz +0.26794919
196	16	15	
197	17	16	
198	18	17	
199	19	18	
200	20	19	
201	21	20	
202	22	21	
203	23	22	
204	24	23	
205	25	24	
206	26	25	
207	27	26	
208	28	27	
209	29	28	
210	30	29	
211	1-Aug	30	Lughnasada +0.60086062
212	2	31	
213	3	1-Aug	
214	4	2	
215	5	3	
216	6	4	
217	7	5	

218	*8*	*6*	
219	*9*	*7*	
220	*10*	*8*	
221	*11*	*9*	
222	*12*	*10*	
223	*13*	*11*	
224	*14*	*12*	
225	*15*	*13*	*Assumption of Mary +1*
226	*16*	*14*	
227	*17*	*15*	
228	*18*	*16*	
229	*19*	*17*	
230	*20*	*18*	
231	*21*	*19*	
232	*22*	*20*	
233	*23*	*21*	
234	*24*	*22*	
235	*25*	*23*	
236	*26*	*24*	
237	*27*	*25*	
238	*28*	*26*	
239	*29*	*27*	
240	*30*	*28*	
241	*1-Sep*	*29*	
242	*2*	*30*	
243	*3*	*31*	
244	*4*	*1-Sep*	
245	*5*	*2*	

246	*6*	*3*	
247	*7*	*4*	
248	*8*	*5*	
249	*9*	*6*	
250	*10*	*7*	
251	*11*	*8*	
252	*12*	*9*	*Solar Magnetic +3.07768354*
253	*13*	*10*	
254	*14*	*11*	
255	*15*	*12*	
256	*16*	*13*	
257	*17*	*14*	
258	*18*	*15*	
259	*19*	*16*	
260	*20*	*17*	
261	*21*	*18*	*Autumnal Equinox +6.31375151*
262	*22*	*19*	
263	*23*	*20*	
264	*24*	*21*	
265	*25*	*22*	
266	*26*	*23*	
267	*27*	*24*	
268	*28*	*25*	
269	*29*	*26*	
270	*30*	*27*	*Radian = Pi+1/2 4.71238898*
271	*1-Oct*	*28*	*Protection Mother of God -57.289962*

272	*2*	*29*	
273	*3*	*30*	
274	*4*	*1-Oct*	
275	*5*	*2*	
276	*6*	*3*	
277	*7*	*4*	
278	*8*	*5*	
279	*9*	*6*	
280	*10*	*7*	
281	*11*	*8*	
282	*12*	*9*	
283	*13*	*10*	
284	*14*	*11*	
285	*15*	*12*	*Harvest Time -3.7320508*
286	*16*	*13*	
287	*17*	*14*	
288	*18*	*15*	
289	*19*	*16*	
290	*20*	*17*	
291	*21*	*18*	
292	*22*	*19*	
293	*23*	*20*	
294	*24*	*21*	
295	*25*	*22*	
296	*26*	*23*	
297	*27*	*24*	
298	*28*	*25*	
299	*29*	*26*	
300	*30*	*27*	

301	*1-Nov*	*28*	*All Saints -1.6642795*
302	*2*	*29*	
303	*3*	*30*	
304	*4*	*31*	
305	*5*	*1-Nov*	
306	*6*	*2*	
307	*7*	*3*	
308	*8*	*4*	
309	*9*	*5*	
310	*10*	*6*	
311	*11*	*7*	
312	*12*	*8*	
313	*13*	*9*	
314	*14*	*10*	
315	*15*	*11*	
316	*16*	*12*	
317	*17*	*13*	
318	*18*	*14*	
319	*19*	*15*	
320	*20*	*16*	
321	*21*	*17*	
322	*22*	*18*	
323	*23*	*19*	
324	*24*	*20*	
325	*25*	*21*	
326	*26*	*22*	
327	*27*	*23*	
328	*28*	*24*	
329	*29*	*25*	

330	*30*	*26*	
331	*1-Dec*	*27*	
332	*2*	*28*	
333	*3*	*29*	
334	*4*	*30*	
335	*5*	*1-Dec*	
336	*6*	*2*	
337	*7*	*3*	
338	*8*	*4*	
339	*9*	*5*	
340	*10*	*6*	
341	*11*	*7*	
342	*12*	*8*	
343	*13*	*9*	
344	*14*	*10*	
345	*15*	*11*	
346	*16*	*12*	
347	*17*	*13*	
348	*18*	*14*	
349	*19*	*15*	
350	*20*	*16*	
351	*21*	*17*	*Winter Solstice -0.1583844*
352	*22*	*18*	
353	*23*	*19*	
354	*24*	*20*	
355	*25*	*21*	*Christmas -0.0874887*
356	*26*	*22*	
357	*27*	*23*	

358	*28*	*24*	
359	*29*	*25*	
360	*30*	*26*	*Radian = Pi doubled 6.28318531*
		27 (+1 day)	
		28 (+2 day)	
		29 (+3 day)	
		30 (+4 day)	
		31 (+5 day)	

Magic Constant of 9

The number 9 has been called the "magic constant," and it shows itself as the regulator of all other numbers. This regulation of numbers controls the golden circle of the sun, and the golden ellipse of our orbit. As we progress in small increments of 9 degrees, we see that the earth reaches the solstices and equinoxes precisely on the nines. When all of the important degrees are reduced to Pythagorean numbers, they are all constantly reduced to the magic nine. The important solar magnetic alignment points of 72° and 252° precede the change from winter to spring, and from summer to fall. All of the major tangent and radian points occur with the magic constant of nine.

Degree	***Pyth Reduction***	***Date***	***Tangent***	***Event***
9	*9*	*Jan 9*	*+0.1583844*	
18	*1+8=9*	*Jan 18*	*+0.3249197*	
27	*2+7=9*	*Jan 27*	*+0.50952545*	
36	*3+6=9*	*Feb 6*	*+0.72654253*	
45	*4+5=9*	*Feb 15*	*+1*	*Lupercalia*
54	*5+4=9*	*Feb 24*	*+1.37638192*	
63	*6+3=9*	*Mar 3*	*+1.96261051*	
72	*7+2=9*	***Mar 12***	***+3.07768354***	***Solar Magnetic Point***
81	*8+1=9*	*Mar 21*	*+6.31375151*	*Vernal Equinox*
90	*9+0=9*	*Mar 30*	***+1.6325E+16***	***Planting***

			Radian 1.57079635	**Season starts 4/1**
99	9+9=18=1+8=9	Apr 9	-6.3137515	
108	1+~~0~~+8=9	Apr 18	-3.0776835	
117	1+1+7=9	Apr 27	-1.9626105	
126	1+2+6=9	May 6	-1.3763819	
135	1+3+5=9	May 15	-1	
144	1+4+4=9	May 24	-0.7265425	
153	1+5+3=9	June 3	-0.5095254	
162	1+6+2=9	June 12	-0.3249197	
171	1+7+1=9	June 21	-0.1583844	Summer Solstice
180	1+8+~~0~~=9	June 30	-1.225E-16 **Radian 3.14159265**	
189	1+8+9=18=1+8=9	July 9	+0.1583844	
198	1+9+8=18=1+8=9	July 18	+0.30573068	
207	2+~~0~~+7=9	July 27	+0.50952545	
216	2+1+6=9	Aug 6	+0.72654253	
225	2+2+5=9	Aug 15	+1	Assumption of Mary
234	2+3+4=9	Aug 24	+1.37638192	
243	2+4+3=9	Sept 3	+1.96261051	
252	2+5+2=9	**Sept 12**	**+3.07768354**	**Solar Magnetic Point**
261	2+6+1=9	Sept 21	+6.31375151	Autumnal Equinox
270	2+7+~~0~~=9	Sept 30	5.4415E+15 **Radian 4.71238898**	
279	2+7+9=18=1+8=9	Oct 9	-6.3137515	
288	2+8+8=18=1+8=9	Oct 18	**-3.0776835**	
297	2+9+7=18=1+8=9	Oct 27	-1.9626105	
306	3+~~0~~+6=9	Nov 6	-1.3763819	
315	3+1+5=9	Nov 15	-1	
324	3+2+4=9	Nov 24	-1.7265425	
333	3+3+3=9	Dec 3	-0.5095254	
342	3+4+2=9	Dec 12	-0.3249197	

351	*3+5+1=9*	*Dec 21*	*-0.1583844*	*Winter Solstice*
360	*3+6+~~0~~=9*	*Dec 30*	*-2.45E-16* ***Radian 6.28318531***	

Pi, Phi, Product, and Square Root

H = Holi-days

Degree	*Pi, Phi, Product, or Sq. Root*	*Date/Diff*
3°	***Perihelion***	***1/3***
51°	*Phi square root*	*2/21 - 48 H*
58°	*Phi*	*2/28 - 7*
60°	*Pi square root*	*2/30 - 2 H*
66°	*Pi/Phi square root*	*3/1 -* ***6*** *H*
72°	***Pi***	***3/12 - 6 H***
78°	*Pi/Phi*	*3/18 -* ***6***
81°	***Vernal Equinox***	***3/21 - 3***
101°	*Pi/Phi*	*4/11 - 20*
107°	*Pi*	*4/17 - 6*
113°	*Pi/Phi square root*	*4/23 - 6 H*
120°	*Pi square root*	*4/30 - 7*
121°	*Phi*	*5/1 - 1 H*
128°	*Phi square root*	*5/8 - 7*
171°	***Summer Solstice***	***6/21 - 43***
184°	***Aphelion***	***7/4 - 13***
231°	*Phi square root*	*8/21 - 103*
238°	*Phi*	*8/28 - 47 H*
240°	*Pi square root*	*8/30 - 2*
246°	*Pi/Phi square root*	*9/6 -* ***6***
252°	***Pi***	***9/12 - 6 H***
258°	*Pi/Phi*	*9/18 -* ***6***
261°	***Autumnal Equinox***	***9/21 - 3***
281°	*Pi/Phi*	*10/11 - 20 H*
287°	*Pi*	*10/17 - 6*

294°	*Pi/Phi square root*	*10/24 - 7*
299°	*Pi square root*	*10/29 - 5*
301°	*Phi*	*11/1 - 2 H*
308°	*Phi square root*	*11/8 - 7 H*
351°	***Winter Solstice***	***12/21 - 43***

Decagon/Dodecahedron of Reality

Plato and Kepler related our physical existence through geometric shapes. The 5 Platonic solids are the tetrahedron (fire), cube (earth), octahedron (air), dodecahedron (ether), and the icosahedron (water). Plato shows how these five forms constitute the construction of our physical reality, the dodecagon and dodecahedron dictates the path of the earth around the sun. The major points on our annual travels around the Sun fall on the 36°, 72°, 108°, 144°, 180°, 216°, 252°, 288°, 324°, and the 360° degrees. The setup of our path was dictated by a 360 day calendar year, and was adjusted to 365 days after a cosmic disaster in the not too distant past. This disaster was noted by civilizations such as the Egyptians, Mayans, Hindus, and by biblical authors, as they attempted to reconcile the changes in the heavens.

The equinoxes and solstices all fall 90 days or degrees apart from each other. The vernal equinox falls on 3/21 at the 81° point, the summer solstice falls on 6/21 at the 171° point, the autumnal equinox falls on 9/21 at the 261° point, and the winter solstice falls on 12/21 at the 351° point.

The equinoxes are preceded by the pi/phi square root, pi, and the pi/phi product in day increments of 6-6-6, and may form one basis for the feared number. The vernal equinox is preceded by the 60° (2/30), 66° (3/6), 72° (3/12), and 78° (3/18) points, before three days later the equinox point arrives at the 81° (3/21) point. This same pattern approaches the autumnal equinox with the 240° (8/30), 246° (9/6), 252° (9/12), and the 268° (9/18) point, with the equinox falling three days later on the 261° (9/21) point for another spacing of 6-6-6.

This insight may be portrayed as pseudoscience, or lacking a base, but science and mathematics bares this understanding out totally. We travel around the sun in an annual ellipse (former circle) that is tracked as circular. This circle that our planet passes through has many tangent and radian points, that has been related in the math systems of pi, phi, pi and phi product, and their associated square roots. While this writer only sees a piece of a magnificent puzzle, it is clear that the decagon points forms the dodecahedron of our physical reality.

Degrees = Days, Months, and Years

Annual Cycle
1 degree = 1 day
360 degrees = 1 year
Precession Cycle
1 degree = 72 years
360 degrees = 25,920 years

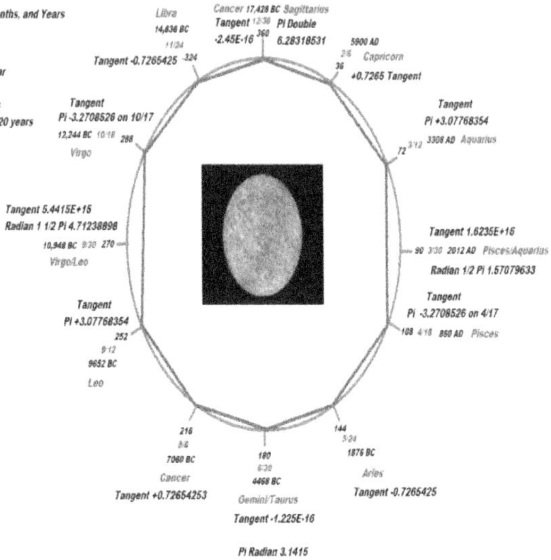

Tangents
Phi = 1.618
58 degree - 2/28 - 3596 AD: +1.60033453
121 degree - 5/1 - 220 AD: -1.6642795
238 degree - 8/28 - 8644 BC: +1.60033453
301 degree - 11/1 - 13,180 BC: -1.6642795

Tangents
Pi = 3.14
72 degree - 3/12 - 3308 AD +3.07768354
107 degree - 4/17 - 788 AD: -3.2708526
252 degree - 9/12 - 9652 BC: +3.07768354
287 degree - 10/17 - 12,172 BC: -3.2708526

Tangents
Pi & Phi Product = 5.0803
78 degree - 3/18 - 2876 AD: +4.70463011
101 degree - 4/11 - 1220 AD: -5.144554
258 degree - 9/19 - 10,084 BC: +4.70463011
281 degree - 10/11 - 11,740 BC: -5.144554

Tangents
Phi Square Root = 1.2720196495
51 degree - 2/21 - 3092 AD: +1.23489716
128 degree - 5/8 - 284 BC: -1.2799416
231 degree - 8/21 - 8140 BC: +1.23849716
308 degree - 11/8 - 13,684 BC: -1.2799416

Tangents
Pi Square Root = 1.77245385091
60 degree - 2/30 - 3452 AD: +1.73205081
120 degree - 4/30 - 868 AD: -1.7320508
240 degree - 8/30 - 8788 BC - +1.73205081
299 degree - 10/29 - 13,036 BC - -1.7320508

Tangents
Pi & Phi Product Square Root = 2.25395208467
66 degree - 3/6 - 3020 AD: +2.24603677
113 degree - 4/23 - 356 AD: -2.3558524
246 degree - 9/6 - 9220 BC: +2.24603677
294 degree - 10/24 - 12,676 BC: -2.3558524

Annual Important Passage Points
3 degree - 1/3 - Perihelion (10/16 in 2560 BC)
81 degree - 3/21 - Vernal Equinox
171 degree - 6/21 - Summer Solstice
184 degree - 7/4 - Aphelion (4/16 in 2560 BC)
261 degree - 9/21 - Autumnal Equinox
351 degree - 12/21 - Winter Solstice

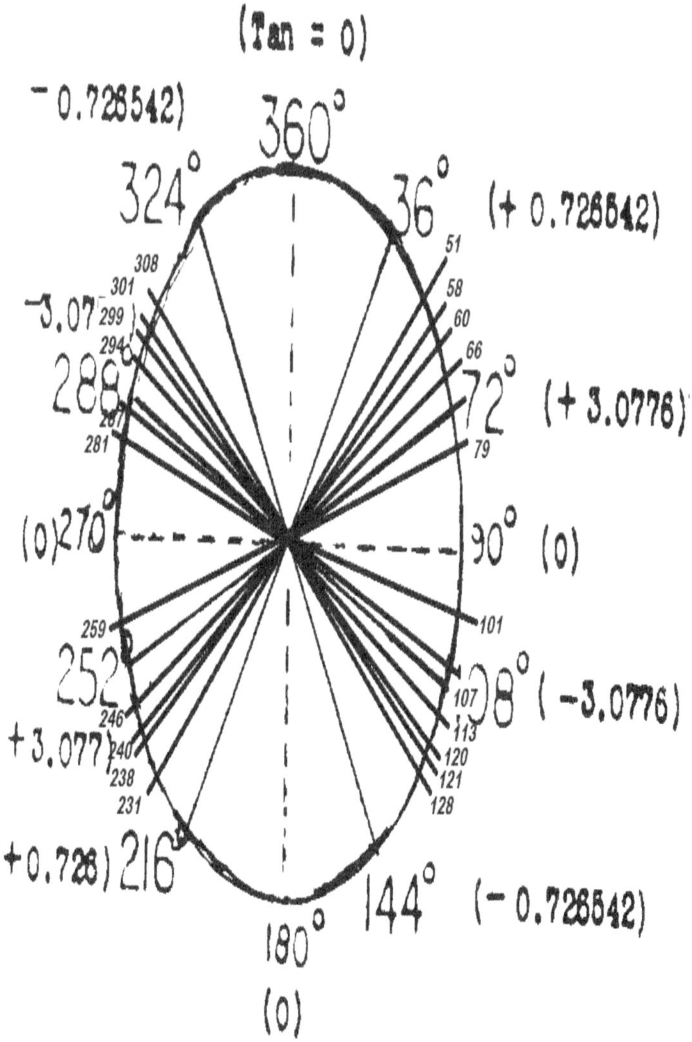

(Tan = 0)
360°
324°
36°
- 0.728542)
(+ 0.728542)
72°
(+ 3.0776)
288°
-3.07
90°
(0)
(0)270°
108°
(-3.0776)
252°
+3.077)
216°
+0.728)
144°
(- 0.728542)
180°
(0)

How does pi, phi, the pi/phi product, and their square roots dictate our seasonal progression around the sun? These systems of geometry based on our physical reality show the relationship between the golden circle, and the golden ellipse of the earth's orbital path. While all points on the 360 degrees of the circle are important, there are certain points that dictate our seasonal journeys.

The first major geometrical point arrives at the 55° point of 2/21 (Parentalia) of phi squared, on the monthly zodiacal transition. We reach the point of phi on 2/28 with the 58° point. Pi squared root is reached on 2/30 at the 60° mark. The 66° of 3/6 (66/6) shows the product of pi and phi square root. The important solar magnetic point of 3/12 arrives at the 72°, as we reach the pi point for the first time on the annual trek. The 79° point of 3/19 (Quinquatrus, Solemnity St. Joseph), shows the product of pi and phi, as we enter spring officially on 3/21. These are all positive tangent points from 1° to 89°.

There is a 22° gap before we reach the first negative tangent point of 101°, on 4/11 (Last day of Passover), as we reach pi and phi product. On 4/17 we achieve the pi point at the 107° tangent. The 113° of 4/23 (Vinalia, Feast of St. George, Yom Ha' atzmaut) equals the product of pi and phi square root. Pi square root is achieved on 4/30 at the 120° mark. Phi is met the next day on 5/1 (May Day/Beltane/Sacrifice to Maia) at the 121° point. Phi square root occurs on 5/8 at the 128° mark. From 91° to 179° are al negative tangent points

We pick up the phi square root mark at the 231° of 8/21 (Consualia), after a 103° gap, as we have passed back in to positive tangent points. The 238° point on 8/28 (Circus Maximus)brings us to a phi point. Pi square root is achieved on 8/30 at the 240° point. The product of pi and phi square root is met on 9/6 (Ludi Romani)at the 246° tangent. The important solar magnetic point of 9/12 is met at the 252° point, as we arrive at a pi tangent. The product of pi and phi is met on 9/19 at the 259° mark, and closes out the positive tangent points. From 181° to 269° are all positive tangent points.

The last negative tangent points begin with the 281° of 10/11 (Meditrinalia), as the product of pi and phi is achieved. Pi is met at the 287° tangent on 10/17. The product of pi and phi squared falls on 10/24 at the 294° point. Phi square root is met on 10/29 at the 299° mark. Phi is met on 11/1 (Temples for Neptune & Pietas, All Souls Day, Nos Gaelan Gaeaf) at the 301° tangent. The square root of phi closes out these points on 11/8 (Tiberius Pater & Gaia, Archangels St. Michael & Gabriel) at the 308° tangent. From 271° to 359°, we find only negative tangent points.

The vernal equinox occurs on the 81° mark of 3/21, and signals the end of the positive tangent points of that side of the circle directly after the pi/phi product point of the 79°, and is 90°'s away from the summer solstice point of 171° on 6/21. The last positive tangent point of 261° on September 21st follows the 259° pi/phi product tangent, and ushers in the autumnal equinox exactly 90°s after the summer solstice.

The original calendar system was based on 360 days, and the 360 degrees of the golden circle. Our present 365 day calendar system was introduced post cosmic disaster of 2344 BCE. The ancients aligned the religious holi-days based on the tangent degrees of the circle, that helped to predict the seasons to come. It is distinctly clear that pi and phi, in our mathematical expressions, is how our physical reality is expressed.

Jesus Christ the Solar Harvest Deity

There has been much debate as to the figure we have come as Jesus Christ. Was he the Son (Sun) of God? Did he perform miracles? Some of these questions may never be answered, but there are some striking correlations to created contrivances for this controversial figure.

One big contrivance is the January 3rd (3°)(+0.05240778 tan) date for the Assumption and Solemnity of Mary (Ch/Ca/LCa), which is simply the perihelion (closest to the sun) of the planet to the sun or lion. This is where the term the 3rd degree is derived from, as the faith was spread through Inquisition.

February 2nd (32°) (+0.62486935 tan) is celebrated as Candlemas (Ca/Ch), and is understand as the day when Christ is presented at the temple. This is again another clear alignment with prior pagan beliefs, as we celebrate Ground Hog Day and Imbolc. The modern usage of an animal to divine the seasonal change, versus the entrails of an animal or a human sacrifice, shows the synthesis of holi-days.

The important solar magnetic point of March 15th (75°)(+3.73205081 tan) again shows the alignment of Jesus Christ (JC) with Julius (JC), as this was the Ides of March, and the day that Julius Caesar was murdered in the Senate. This day is Purim on the Jewish calendar, as the day that the walls of Jerusalem were breached by the Babylonians, yet the Jewish calendar is based on the Babylonian calendar. We also reach the Pi/Phi product square root at this degree tangent and Pi (1.30899694 radian). March 19th (79°)(+5.14455402 tan) marks the Solemnity of St. Joseph (Ca), celebrating the vernal equinox on 3/21, and meets the Pi/Phi product.

The traditional beginning date for planting season begins on April 1st (91°)(-28.636253 tan), as this is celebrated as The Feast of Kings (Ch/Ca) and marks the onset of the Easter (Eostara) season, which has its roots in the pagan holi-day system. We also celebrate this day as April Fools Day (US), as you were considered an April Fool for believing in the Feast of Kings! We pass the Pi radian of 1.60570291 on April 2nd.

April 15th (105°)(-3.7320508 tan) is celebrated as the Crucifixion of Christ (Ch/Ca), and is served as Tax Day in the United States, as hopefully the tax man won't crucify you, but will gently pass over your house! The former aphelion of 2560 BCE occurred the next day on 4/16.

October 15th (285°)(-3.7320508 tan) is celebrated in Catholicism as St. Teresa of Avila day, but this is clearly another pagan tie in, as this is the traditional onset of harvest time seasonally. The Immaculate Conception of Christ (Ca) is celebrated on December 8th (-0.404262 tan). Again the human gestation period is 280 days, and she would have given birth on September 18th at the 258° point, just prior to the autumnal equinox on September 21st. This is again another curious seasonal correlation.

December 25th (355°)(-0.0874887 tan) is celebrated as Christmas (Ch), Nativity of Christ (Ca/LCa), and is the day when the risen son (sun) climbs one degree higher on the horizon, as it signals the onset of warmer weather to come. Over 30 deities that predate the birth of Jesus Christ holds this day as their date of birth, such as; Horus, Tammuz, Krishna, Mithras, Dionysius, Apollo, and many more. This day was also Dies Natalis Sol Invictus in Rome (Ca), and means Birth of the Unconquered Sun (Son). It is startlingly clear that Jesus Christ was aligned as a Solar Harvest Deity!

355°(12/25) - 280 days = 75°(3/15) or Conception!

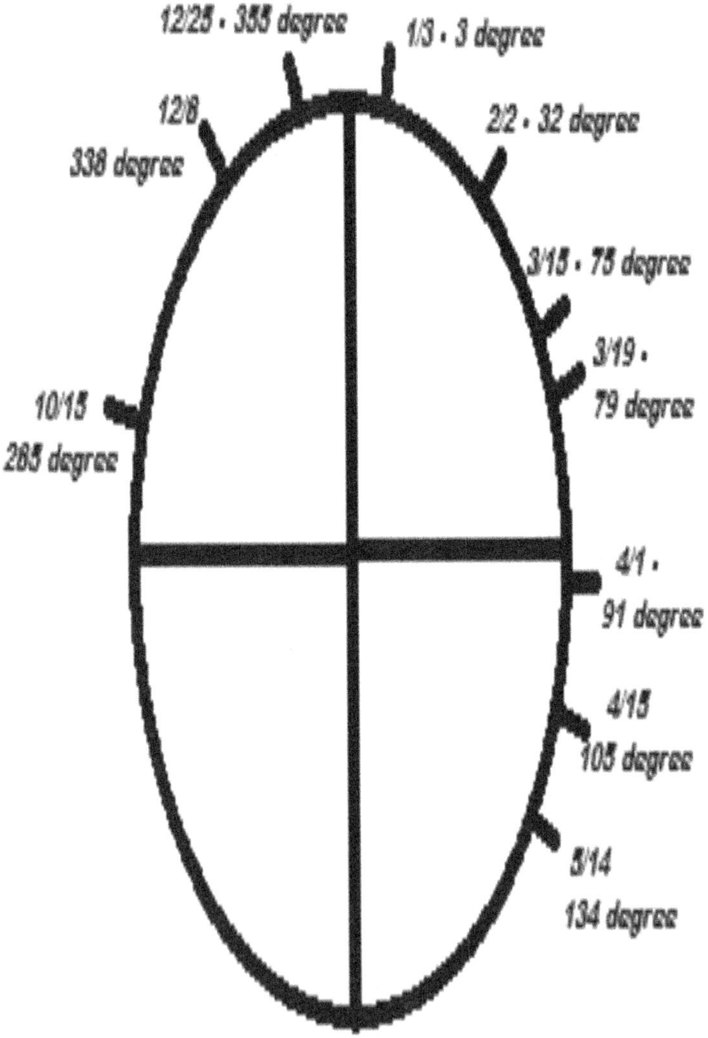

Jesus Christ Solar Harvest Deity

Printed in Dunstable, United Kingdom

91218908R00117